S0-BDW-839

Sustainable Development Challenges in the Arab States of the Gulf

The Gulf Research Center Book Series at Gerlach Press

The GCC in the Global Economy
Ed. by Richard Youngs
ISBN 9783940924018, 2012

Resources Blessed: Diversification and the Gulf Development Model
Ed. by Giacomo Luciani
ISBN 9783940924025, 2012

GCC Financial Markets: The World's New Money Centers
Ed. by Eckart Woertz
ISBN 9783940924032, 2012

National Employment, Migration and Education in the GCC
Ed. by Steffen Hertog
ISBN 9783940924049, 2012

Asia-Gulf Economic Relations in the 21st Century The Local to Global Transformation
Ed. by Tim Niblock with Monica Malik
ISBN 9783940924100, 2013

A New Gulf Security Architecture Prospects and Challenges for an Asian Role
Ed. by Ranjit Gupta, Abubaker Bagader, Talmiz Ahmad and N. Janardhan
ISBN 9783940924360, 2014

Gulf Charities and Islamic Philanthropy in the 'Age of Terror' and Beyond
Ed. by Robert Lacey and Jonathan Benthall
ISBN 9783940924322, 2014

State-Society Relations in the Arab Gulf States
Ed. by Mazhar Al-Zoby and Birol Baskan
ISBN 9783940924384, 2014

Gulf Research Centre Cambridge
Knowledge for All

Political Economy of Energy Reform The Clean Energy-Fossil Fuel Balance in the Gulf
Ed. by Giacomo Luciani and Rabia Ferroukhi
ISBN 9783940924407, 2014

Security Dynamics of East Asia in the Gulf Region
Ed. by Tim Niblock with Yang Guang
ISBN 9783940924483, 2014

Islamic Finance Political Economy, Performance and Risk
Ed. by Mehmet Asutay and Abdullah Turkistani
ISBN 9783940924124, 3 vols set, 2015

Employment and Career Motivation in the Arab Gulf States: The Rentier Mentality Revisited
Ed. by Annika Kropf and Mohamed Ramady
ISBN 9783940924605, 2015

The Changing Energy Landscape in the Gulf: Strategic Implications
Ed. by Gawdat Bahgat
ISBN 9783940924643, 2015

Sustainable Development Challenges in the Arab Sates of the Gulf
Ed. by David Bryde, Yusra Mouzughi and Turki Al Rasheed
ISBN 9783940924629, 2015

The United States and the Gulf: Shifting Pressures, Strategies and Alignments
Ed. by Steven W. Hook and Tim Niblock
ISBN 9783940924667, 2015

Africa and the Gulf Region: Blurred Boundaries and Shifting Ties
Ed. by Rogaia Mustafa Abusharaf and Dale F. Eickelman
ISBN 9783940924704, 2015

Rebuilding Yemen: Political, Economic and Social Challenges
Ed. by Noel Brehony and Saud Al-Sarhan
ISBN 9783940924681, 2015

Sustainable Development Challenges in the Arab States of the Gulf

Edited by David Bryde, Yusra Mouzughi and Turki Faisal Al Rasheed

First published 2015
by Gerlach Press
Berlin, Germany
www.gerlach-press.de

Cover Design: www.brandnewdesign.de, Hamburg
Printed and bound in Germany by Hubert & Co., Göttingen

British Library Cataloguing in Publication Data.
A catalogue record for this book is available from the British Library.

Bibliographic data available from Deutsche Nationalbibliothek
http://d-nb.info/1070508888

ISBN: 978-3-940924-62-9 (hardcover)
ISBN: 978-3-940924-63-6 (ebook)

Acknowledgements

We are delighted and honored to have edited Sustainable Development Challenges in the Arab States of the Gulf and appreciate the generosity of the people who have given us their support.

First of all, we are extremely grateful to the attendees of the Gulf Research Meetings in Cambridge, UK, who participated in our workshops *Sustainable Development Challenges in the GCC,* in 2013, and *Addressing the Sustainability Agenda in the Gulf Region,* in 2014, who have given their trust and willingness to publish their work and experiences; without them this book would not be available.

Secondly, our debt and gratitude are owed to our co-editors of this book for their patience and perseverance during its compilation.

Thirdly, special thanks to Mr Ronnie De Guzman and Ronnie Arevalo from the beginning to the end of the book journey; they tirelessly gave their helping hand to make this book come to reality.

Fourthly, special thanks to Gerlach Press for helping us through the process of writing, editing, and supervising this book.

Finally, editors David Bryde, Yusra Mouzughi, and Turki Faisal Al Rasheed feel lucky and blessed that we have collaborated on this memorable book; it cemented and made our friendship stronger than ever. Furthermore, our families, friends and fellow academics have given their trust and unwavering support for the completion of this book.

With this, we as editors believe the work and experience in publishing this book will be useful to all stakeholders, particularly the decision makers, professionals, academics, postgraduate students and undergraduate students in the GCC in particular, and in the rest of the Arab World and the World in general, to achieve GCC sustainable development goals.

Dedication

Sustainable Development Challenges in the Arab States of the Gulf is dedicated to our families, who have provided their unconditional love and support 'despite our heavy commitments, which took us away from our families, social obligations and professional duties while we were undertaking the challenging task of writing this book The motivation from our families has inspired and encouraged us every step of the way to complete the journey in finishing this book.

Moreover, Sustainable Development Challenges in the Arab States of the Gulf is dedicated to the United Nations Sustainable Development Goals Open Working Group, which is implementing sustainability – particularly to all the stakeholders of the sustainability challenges in the Gulf Cooperation Council (GCC) countries and the Arab World in general who strive to improve the triple bottom line of economic, social and environmental sustainability for the advancement of their societies.

Contents

Introduction

David Bryde, Yura Mouzughi and Turki Faisal Als Rasheed

By way of providing an introduction to this book, which contains edited chapters on the topic of sustainable development, it is useful to give a sense of the landscape of the topic as it applies to the Gulf Co-Operation Council (GCC) countries. The overall rationale for the book is that there are generic sustainability needs and concerns to be addressed that encompass all of the GCC i.e. all countries face challenges in terms of sustainable consumption and production. Some of the chapters in the book focus on generic issues – see for example Chapter 1.

Many of these needs and concerns are increasing in importance for the GCC. For example, in respect of environmental sustainability, some GCC countries are facing energy deficits and will be net importers of energy in the near future; and hence are considering the use of more sustainable sources of energy such as renewables. In relation to economic sustainability, the Middle East region has one of the highest youth unemployment rates in the world and there is a great challenge in developing a skilled workforce and providing them with jobs. The fact that the GCC countries are the largest recipients of temporary migrants in the world has implications for social sustainability in terms of the coherence and fabric of societies. In general, within the Middle East and North Africa (MENA) regions, the failure of countries to engage in sustainable development, with particular consideration of the social and economic impacts, contributed to the Arab Spring.

Indeed it has long been recognised that there are general sustainability-related issues facing the countries in the Middle East. For example, if one goes back over 20 years to the international framework treaty signed at the 1992 United Nations Framework Convention on Climate Change (UNFCCC) the 6 member countries of the GCC (Bahrain, Kuwait, Oman, Saudi Arabia, United Arab Emirates and Qatar) were individually included in the treaty as one of 154 "Non-Annex I" parties. [The UNFCCC was a "Rio Convention", one of three adopted at the "Rio Earth Summit" held in Rio de Janeiro in 1992 (UNCED, 1992), which is widely regarded as a landmark summit in terms of putting the topic of sustainable development on the global agenda.]

For example, the Non-Annex I parties were regarded by the UNFCCC as being particularly vulnerable to the adverse impacts of climate change i.e. being prone to desertification and drought or the potential economic impacts of climate change response measures i.e. due to being heavily reliant on income from fossil fuel production. Such countries, the Convention stated, needed the promotion of activities within the treaty which answered their special needs and concerns, such as investment, insurance and technology transfer (UNFCCC, 1992).

It is also in the Convention's spirit of proposing the taking of technologies utilized in one part of the world and applying them to countries in other regions that we include a chapter on methods and tools to understand the complexity of food security that have been developed in the context of rural Malawi (see Chapter 10).

However whilst acknowledging the common sustainability agenda across the GCC we also recognize that in terms of sustainable development individual GCC countries face their own unique contextual challenges at a local, regional and national level. This is reflected in the fact that some of the chapter contributors to the book focus their attention on a particular GCC country i.e. water security in the UAE (see the example in Chapter 8). Of course water security is of interest to all GCC countries, though the specific issues may vary e.g. in Saudi Arabia a specific challenge has been balancing the need for water against its strategies for self-sufficiency in certain food products.

The GCC countries are increasingly being faced by the challenges that have confronted western countries in terms of how to balance the need to develop their economies whilst at the same time taking into account the impact of such developments on the environment and on communities and individuals; the classic concepts of "profit" v "planet" v "people" or balancing the Triple Bottom Line (TBL) of "environmental", "economic" and "social" sustainability (Elkington, 1997).

As the GCC develop their built environments - including commercial and non-commercial properties and their infrastructures; their industrial capacities - often through large-scale industrialization programs; and their agricultural capacities, there are often inter-linked sustainability issues to consider. Taking the three sustainability pillars of the TBL in turn, there are: 1) environmental-sustainability issues such as: waste, recycling, water usage, energy – including the use of renewables, and pollution – to name but a few; 2) economic-sustainability issues including: employment opportunities for local people, education and training and engagement of business and individuals that make up the supply chains; 3) social-sustainability issues such as: safety at work, working hours, equality and diversity, noise dust and pollution, traffic congestion, stakeholder engagement and community involvement in decision-making. Undertaking activities to address each pillar individually is challenging enough, but it adds an additional layer of complexity when one considers that an activity in one pillar area can often involve trade-offs and tensions in another pillar.

There is often a perception in the west that countries in the GCC are playing catch up in terms of sustainable development. That the rapid urbanization and modernization of cities, as evidenced by the development of real estate, is taking place with little regards for sustainability. Yet the picture is more nuanced than that. For example, whilst traditionally the GCC countries have followed the lead taken by Western countries there are pockets of excellence in which individual countries are taking their own initiative. A case in point is the development of environmentally-friendly or "green" buildings. A typical response in the GCC has been to take standards developed in the west i.e. the LEED or BREEAM accreditation systems from the US and UK respectively and ensure new buildings meet their requirements. But some countries are going beyond this and developing their own accreditation systems i.e. the PEARL system in Abu Dhabi and the QSAS in Qatar. So in this one small area of sustainability there is evidence of heterogeneity and initiatives and activities that reflects local needs and concerns. There are many other examples encompassing environmental, economic and social sustainability where policies and practices in the GCC member countries are not happening in an homogeneous way.

So with this sustainable development landscape in mind, in late 2012 we formulated a proposal to run a workshop at the Gulf Research Meeting (GRM) Annual Conference of 2013 at the University of Cambridge, UK on the topic of "Sustainable Development Challenges in the GCC". The aim of the workshop was to provide a forum for people interested in the broad topic of sustainability from both inside and outside the GCC to engage in discussion and debate related to the key challenges the GCC face in ensuring that developments in their countries take place in a sustainable fashion.

The workshop was timely, as it was our view that up to that point much of the research in the area of sustainability had been western-centric and there was a need to broaden our understanding of the subject to other contexts, such as the Middle East. By reviewing and discussing aspects of sustainability in the GCC countries we also saw the potential to contribute to the expansion of Gulf Studies, with sustainable development being increasingly important if one wishes to gain an holistic and comprehensive understanding of what is happening in the region. It was also our belief that the outcomes from the workshop would help decision-makers in both the public and the private sector organisations to better understand and develop appropriate policies and practices for sustainable development in their countries.

We sought contributions from all those with an interest in sustainable development in the GCC, encompassing academics and those involved in policy-making and practice; and including many different perspectives i.e. science, engineering, politics, economics, social sciences and management sciences. Contributions could take various forms, including: case studies, surveys, reflective studies, conceptual papers, and policy and practice statements. It was our hope that we would attract papers from individuals working in the following organisations:

- Government and public agencies responsible for sustainability-related policy and regulation
- National and international companies involved in sustainability-related activities in the GCC

- Regional institutions promoting sustainable development
- International accreditation bodies
- Academic and other research institutions

The workshop was subsequently held in July 2013 and the outcome more than met expectations. Participants based in UAE, Bahrain, Lebanon, Egypt, UK and US presented their papers across two days under the broad themes of: consumption and production; urban development, planning and real estate; climate change, environment and energy; water management; role of the public and private sectors; and knowledge and awareness. After the success of the workshop we decided to build on the momentum achieved and hold a follow-on doctoral symposium at GRM 2014. Here the intention was to provide a forum to engage new researchers that were studying aspects of sustainability which either directly related to, or would be of interest to, the GCC countries. The symposium was held at GRM 2014, University of Cambridge, in August 2014. As with the workshop in 2013, the symposium proved to be a highly interactive and thought-provoking two days with papers presented by early career researchers studying in Oman, Bahrain, Saudi Arabia, Morocco, Qatar and the UK. Topics covered included: bridging the gaps between policies and practices, issues related to awareness, education and knowledge transfer, methods to minimize waste and carbon footprints, food security, the role of foreign direct investment, renewable energy and rare sources of water.

The breadth and variety of topics presented by the participants at both GRM 2013 and GRM 2014 reflects the importance of taking a multi-disciplinary perspective when viewing the subject of sustainable development. This perspective encompasses those working in the disciplines of environmental sciences, engineering and technology, public policy and social sciences. Taking a multi-disciplinary perspective is crucial as academics, policy makers and practitioners seek a deeper and an holistic understanding of the complexities of the topic; and look to provide solutions to the challenges facing countries in the GCC as they seek to address and balance the often competing demands of the TBL or *People, Planet* and *Profit*. In selecting chapters for inclusion in this book, the need to reflect such a multi-disciplinary perspective was upmost in our minds.

In **Chapter 1 Kassem El-Saddik** analyses the opportunities and challenges facing the GCC countries as they seek to move towards more sustainable consumption and production economic models. The chapter explains how economic development of GCC states over the last three decades has been achieved at unprecedented rates but at incalculable cost to the region's scarce natural and environmental resources. Besides the future need for the importing of oil and gas, due to increasing internal demands, the GCC are consuming growing quantities of their underground reserves to fuel their rapid economic and urbanization growth. Demand for electricity is soaring and is projected to grow by seven to eight percent annually for the next decade in order to meet the escalating energy required to meet basic needs of air-conditioning and water. With the water produced primarily from energy-intensive desalination processes. So resolving the tensions between their socio-economic development and their inefficient energy use, which often has damaging environmental impacts, becomes an inevitable challenge that GCC states are facing. As the chapter goes on to explain this challenge will more than likely be exacerbated in the face of the growing negative consequences of climate change and the growing costs associated with mitigating for the scarcity of water resources in the region. Indeed, as the chapter details, in response to some of these challenges there have been recent initiatives towards addressing sustainability in the energy and water sectors, as affirmed in the 2009 Arab Regional Strategy for Sustainable Consumption and Production (ARSSCP, 2009). Informed by the overall Sustainable Consumption and Production (SCP) framework adopted in the Marrakech Process (UNEP, 2014) El-Saddik derives a GCC-specific SCP framework. A key element of the framework is the GCC context of the "Rentier State", which is the underpinning element that explains policies and legislation influencing production and consumption. In a Rentier State model the "rents" the country receives from oil revenues, which far outweigh the costs of production, funds a generous welfare state and enables the government to provide citizens with subsidised or free goods and services. Such a model is seen to encourage high and increasingly unsustainable levels of consumption and to inhibit private sector growth and innovation. The chapter uses the author's ARSSCP framework to question the key issues related to the region's SCP priorities, to appraise the progress achieved since the ARSSCP, to explore opportunities and challenges, and finally to prescribes policy options and drivers for a successful transition to more sustainable consumption and production patterns.

In **Chapter 2** by **Mhamed Biygautane and Justin Dargin** we move from the GCC as a whole to consider the specific context of the United Arab Emirates (UAE). Here the authors analyse the economic, environmental and social achievements and challenges as the UAE search for ways to achieve more sustainable development. The chapter provides a historical background to sustainability in the UAE. It explains how the country's success comes in part from its ability to emphasize two core parts of its identity: firstly, the capacity to reinvent itself; and secondly its resilience to financial distress. Drawing on its roots as a cross-cultural *entrepôt*, the UAE was forced to adapt and rebuild its economy after World War II as a trading-hub based on re-exports. With the discovery of oil in 1966, Dubai and Abu Dhabi together led the founding of the United Arab Emirates in 1971. The most recent period of growth and transformation of the UAE's economy is at once a break from the past and yet consistent with this cultural record of reinvention, with sustainability of its economy, society and environment a part of this culture. The chapter explains how the UAE has constantly demonstrated its commitment to raise environmental awareness of sustainability and, in the words of its Vision 2021 Strategic Document "to implement innovative solutions to protect and sustain the environment using new, energy-efficient technologies to reduce its carbon footprint"

(UAE Ministry of Cabinet Affairs, 2014). The UAE government's motives are directed towards making both Emirati citizens and other residents more responsible and aware of the nation's ecological deficit by promoting environmental awareness, mainly through preventive measures i.e. carbon dioxide emission reduction and regulation to protect ecosystems from urban development. Biygautane and Dargin explain how the UAE has achieved remarkable success in a short period of time, against considerable obstacles. As they argue, the fact that the country faces economic, environmental, social and cultural challenges is no surprise. But as they further argue, it offers valuable lessons to the GCC in general as it continues to address these challenges and in the process attract human talent and investments from all around the world.

Chapter 3 by **Paul Joyce** explores the relationships between environmental performance and public governance in the GCC. It seeks to answer the following broad research question: Is public governance a significant factor in variations in environmental performance among the GCC member states? To answer this question a methodology was adopted which comprised of a comparative analysis of the trends and developments of the six GCC states using secondary data sources. The primary source of data is the World Development Indicators from the World Bank. Joyce's analysis, which covers the eleven year period from 2000-2011, reveals a number of trends and developments in public governance that provoke further questions. Following a theme developed in Chapter 1, the analysis is placed in the context of the Rentier State model and whether there are indicators that the GCC countries are moving away from it. For example, the findings on the perceptions of the extent to which citizens are able to take part in selecting their government and also on perceptions of the presence of some liberal democratic freedoms suggests either a problem for the modernisation of the governance of the GCC states or shows that the GCC states are following their own path to modernisation. As acknowledged in the chapter some trends and developments found in the data are puzzling. For example, the records on economic growth and environmental performance are somewhat at odds with each other. A case in point that was discussed in the chapter is the case of the UAE. Specifically the evidence of the highly rated competence of the country using the World Bank's "Government Effectiveness" indicator and its record on environmental improvement (which the data showed was second to none among the GCC states) seems logical and consistent with other sources of data. But by the same token, why according to the same data source had the country suffered a massive drop in terms of GDP growth (on per capita basis) after 2004? Did this drop occur despite very effective public governance institutions? **Joyce** furnishes the chapter with a discussion of these and other noteworthy findings. He raises the possibility that traditional capabilities used by bureaucratic governments could be important for national improvements in environmental performance before ending with the proposition that in the future it is possible that strategic-state capabilities will matter more for government effectiveness in delivering better government performance in meeting environmental sustainability targets than is currently the case.

The theme of governance is continued in **Chapter 4** in which **Jerry Kolo** sets out the imperative of and a framework for environmental governance. This chapter contends that, individually and collectively, the member states of the GCC demonstrate political will and commit resources, along with numerous policy and programmatic initiatives, to address the region's myriad of complex and intertwined environmental challenges. Yet, ample evidence and data exist to show that the region faces critical and urgent environmental challenges, which have a direct bearing on the region's prosperity and future aspirations. This chapter sides with the school of thought that a critically important but either missing or inadequate dimension of sustainable development in the GCC is environmental governance. With this underlying premise the chapter examines environmental governance in the GCC. It identifies the obstacles to its effective implementation and, where it exists, its weaknesses. Further it proposes a Collaborative Environment Governance Framework for Sustainable Development and Environmental Protection in the region. Using a desktop research method, the chapter provides data from two secondary sources to achieve its aims. The first source consists of professional and scholarly environmental publications focusing on the GCC, and the second is content from major print media in the region. Information gleaned from these sources is used to review and articulate the concept of environmental governance, and construct the Collaborative Environment Governance Framework for Sustainable Development and Environmental Protection. The framework identifies the roles and responsibilities of the main environmental governance stakeholders/actors in the three categories of GCC/Regional, States/National and non-state. For each of the three categories the framework sets out roles and responsibilities in terms of seven "imperatives": 1) policy making, 2) legislation, 3) budgets, 4) administration and management 5) inter-governmental coordination and public-private partnerships, 6) research, education and training and 7) community outreach and citizen mobilization. Kolo suggests that scaling up on environmental governance in the GCC would be facilitated by exploiting the broad geo-spatial, environmental, religious, economic, cultural and even political similarities of most or all the countries in the region. In this sense, coordinated and integrated efforts through environmental governance to address the environmental challenges in the region would be in the best interest of the member States and of future generations.

The role of the private sector in GCC countries achieving inclusive and sustainable growth is the topic covered by **Yousuf Hamad Al-Balushi** in **Chapter 5**. The private sector is an important part of any given economy and it is an essential component to boost the growth and stability of the world economy. Overall, GCC countries have been performing quite well in the last four decades using a state-led development model. However, a number of social and economic sustainability challenges lie ahead, including the creation of adequate employment opportunities for nationals, sustaining a high economic growth-rate and securing funding for future development projects. The private sector can significantly contribute to meeting and addressing these challenges, primarily through reducing the reliance on oil revenues. Such a reduction in reliance on oil revenue will require sectoral diversification and national human capital formation. A significant body of literature has developed that deals with the private sector as an engine of growth, especially in developed economies such as the United States, Europe, Japan etc. and to some degree in a number of emerging economies i.e. in Latin American and Asian countries. In contrast, little is known about the role of the private sector as an engine of growth to meet sustainable development challenges in oil-producing countries such as those of the GCC. These countries seem to have weak and somewhat inefficient private sectors in terms of economic diversification, capital formation and employment generation for nationals, and largely depend on the government – again the concept of the "Rentier State" looms large. The chapter provides an analysis of the whole set of complex factors limiting private-led growth in the Gulf region. Al-Balushi does this by considering four broad themes: firstly, the need for development of the GCC countries to be seen in the context of a highly connected and inter-dependent global economy; secondly, the specific challenges, with an emphasis on the labour market structures and issues associated with national identity; thirdly, the institutional frameworks needed to support and enhance the private sector; and fourthly, a set of policy recommendations to address some of the constraints and inhibitors to an enhanced role for the private sector in economic development in the GCC countries.

The theme of awareness is continued in **Chapter 6** by **Nilly Kamal Elamir.** This chapter focuses on some of the environmental risks to achieving sustainable development and the need to incorporate a clear Disaster Risk Management (DRM) system within any proposed development model. Typical negative risks are oil spills and their adverse impact on the natural environment. The chapter highlights the important role that awareness plays in mitigating for the threats of such events [in the worst case – disasters], with awareness of the potential adverse consequences of certain actions on the environment being an integral part of an effective DRM. It is posited that a crucial step in managing the environment is having systems in place to link the natural characteristics present with people's understanding of those characteristics. So an effective DRM will clearly identify the salient environmental characteristics and the corresponding risks of certain if certain actions are undertaken or events occur. The next step is to ensure this information is effectively communicated to all the people that either might carry out the activities or instigate [or be affected by] the events. The chapter presents an argument that high levels of awareness are an outcome of three key stakeholder groups playing an active role in furthering understanding of sustainable development principles and practices. These stakeholder groups are: the individual residents in the GCC countries [including the societal interactions between individuals], scientists working in public and private sector organisations in the countries and, finally, the policy makers. **Elamir** uses the perspectives of these three stakeholder groups to analyse levels of awareness at the science and technology level, the individual and civil society level and the policy action level. The chapter stresses the importance of reflecting both the global perspective of a desire to achieve sustainable consumption and production and also the local GCC perspective; which in the case of the latter means that education to raise environmental awareness can build upon traditional Islamic values which align well with the ethos of sustainable development.

The Gulf economic model has shown development comes as a top priority where sustainability sometimes comes as second priority. During the last decades, huge amounts of cash flow have been injected into the infrastructure sector in Gulf region. These mega projects have definitely affected and even changed the environmental and natural characteristics of the Gulf States. While policy makers in the Gulf have been adopting very ambitious development plans for moving their countries from the developing block to the developed one; the policy trends seem to lack strong environmental and Disaster Risk Management (DRM) dimensions. Since modern science and applied science have contributed a lot to mitigation of the environmental crises which includes DRM literature, this chapter looks at the research finding of DRM for a better environment status in the Gulf. The chapter intends to introduce practical solutions and policy recommendations. This chapter explains the types of natural and environmental threats in the region by looking at environmental crises in the Gulf, then the chapter will examine how these threats are considered on the awareness level and to what extend the DRM is used in the Gulf to combat such challenges. Also, the chapter will deal with recommendations towards a more balanced and Eco-system in the region using the DRM tools for sustainable development in the Gulf. As environment is cross border, the chapter includes examples from the Gulf Cooperation Council (GCC) as well as other Arab, Asian or Middle Eastern countries.

Chapter 7 by **Latifa Al-Khalifa** uses the case of the Kingdom of Bahrain to explore how sustainable development can be advanced using what the author refers to as a "triangular approach" (TA). In essence the TA approach involves key stakeholder groups coalescing in an iterative process of planning, monitoring and evaluation. An integral part to the TA is the inclusion of cohesive elements such as the integration of ICT, the development of a research culture, and the mainstreaming of the national goals of sustainability to the smallest units of society. In the example described in the chapter this coalescence has the ultimate goal of promoting sustainability awareness literacy, which is a necessary precursor to advancing sustainable development practices. The primary emphasis in terms of key stakeholders is on the roles of 1) youth and women 2) government and 3) the private sector. The development of such a framework is timely and necessary as a response to the need to provide a route for development and growth that is rooted on the premise of positive conservation, protection, and continuity. In the first part of the chapter Al-Khalifa focuses on the lack of awareness among the stakeholders on the basic tenets of sustainable development. A term called Sustainable Development Awareness (SDA) is introduced and described exhaustively in this section of the chapter using Bahrain as the local environment. Formulation of this construct is based on the conceptualizations available in both theoretical and empirical literature. The second section of the chapter builds a discourse that centres on the specific actions that are needed to be undertaken in order to address the current status of Bahrain in terms of its low levels of SDA. The last section of the chapter offers a discussion on the anticipated challenges in successfully applying the TA in the context of Bahrain. High on the list of challenges are the threats of socio-cultural conflicts and competing economic demands. As well as ensuring effective governance is in place and top-down policy formulation and implementation from government in terms of activities to raise awareness of sustainable development an holistic and optimized approach needs to harness the inputs and insights from youth and women and the private sector.

The next three chapters address a subject that is high on the agenda of the GCC countries namely, water security and food security. Firstly in **Chapter 8 Rachael McDonnell** focuses on the water security stresses that challenge sustainable development in the UAE. Arguably, of all the natural resource challenges facing the GCC countries, water scarcity and with it water security is arguably the most fundamental problem to be overcome. The declining strategic reserves of groundwater allied with growing salinization of both inland and coastal aquifers is leading to declining and degraded water resources. In the UAE the estimated groundwater reserves and desalinated water stores highlight the stresses in the resource system. In this chapter McDonnell analyses the complex challenges to managing water wisely from both a resource as well as institutional and legal standpoints. The UAE constitution states that water is the property of each Emirate and the resulting institutional fragmentation allied with limited alignment of policies across sectors has ensured the natural system is being unintentionally stressed. The different legal and regulatory frameworks bring varying controls on the water service and resource sectors. Given the shared aquifers, shared sea and shared air this can bring difficulties in managing the environmental impact of the water industries. Aligning the economic development plans for the UAE against the natural resource base is an ongoing challenge with many more steps needed before this is achieved. As the chapter explains the changing relationship between society and water is one of the most complex and intractable to understand and manage. With the provision of tapped and bottled water to citizens and communities the traditional relationship between desert dwellers and this most precious of resources has been changed beyond all recognition. The disconnection between nature and society and the transfer of risk management and responsibility to governments challenges future water security and so sustainable development possibilities. A solution to this challenge is for individual citizens to take back some responsibility for water production and distribution to ensure that their demands for water are met, whilst at the same time such demands do not lead to systems of supply that have adverse effects on the socio-economic and natural environments in which they live.

Chapter 9 by **John Anthony Allan, Mark Mulligan and Martin Keulertz** focuses on the food and water supply chains in the GCC. Alongside water security ensuring a secure supply of food is one of the most urgent issues to address to achieve long-term economic sustainability in the GCC economies. Due to the lack of water resources readily available for food production, the GCC economies need to find alternative solutions in the form of food-water imports through food supply chains. However, as this chapter shows the GCC economies are dependent on the international market and thus on global food-water.

Allan, Mulligan and Keulertz illustrate the role of water in food supply chains and how the GCC economies can sustainably steward food-water. This can be done by taking a supply chain approach by first, managing food-water carefully through waste-reducing policies and second, through the sustainable and responsible investment in land and water resources in overseas territories. Through a case study scenario of Sudan the chapter highlights how previous and supposedly ongoing plans to import food-water from East Africa prove to be a risky and unsustainable strategy. Hence alternatives are needed and the chapter analyses alternatives in food-water imports from Romania, where land and water resources are available to increase production in a more sustainable manner. The authors' analysis of the cases of Sudan and Romanis identify three key bio-physical risks when considering areas to focus foreign direct investment (FDI) relating to food production, which are: baseline aridity, prior vegetation cover and proximity to population. The chapter ends with a few recommendations on how to sustainably steward supply chain management by placing food-water at the heart of initiatives of GCC food and water security plans. Good stewardship recognises that food water at the start of a supply chain is particularly at risk where virgin land is transformed for agricultural production without properly accounting for the needs of the local populations and the impacts on their livelihoods. So, for example, if the land is selected on the basis of the characteristic of the prior vegetation i.e. replacement of water-intensive cotton or lucerne crops by less water-intensive wheat cultivation the transformation of the land can create win-win situations for the indigenous populations and the GCC countries at the end of the supply chain.

Chapter 10 by **Samantha Dobbie, James G. Dyke and Kate Schreckenberg** introduce us to the use of participatory methods and simulations to understanding the complexity of rural food security. The realisation of food security is hindered by its inherently complex nature. Within less developed countries, a number of contributing factors act to undermine food security. At the level of rural households, livelihoods remain largely dependent upon agriculture. Structural factors such as small plot sizes, widespread soil degradation and a continued dependence upon rain fed agriculture, leave smallholders vulnerable to climatic shocks. Dobbie, Dyke and Schreckenberg describe field work undertaken in Malawi to explore the potential role of simulation tools in understanding the complexity of sustainable food provision. The chapter sets out how the use of simulation techniques can act as a tool to explore the complex social, ecological and political factors affecting food availability, access and utilisation. Agent-based modelling (ABM) is one possible technique, which comprises a computerised simulation of agents located within an environment. Each agent provides a computational representation of a real-world actor, such as an individual, household or institution. Behavioural decisions and coping strategies of farmers in the face of drought formed the focus of the field study. A participatory rural appraisal (PRA) exercise was carried out which was designed to elicit greater understanding of smallholder responses to drought; as well as the perceived impact of government interventions in the form of input subsidies. PRA is a broad term that encompasses a wide range of data collection techniques that aim to involve local stakeholders in analysing their own situation. Tools include matrix scoring, seasonal calendars and wellbeing ranking exercises. Results from participatory fieldwork were successfully incorporated into an agent-based model of Malawian smallholders. Initial implementation of the model found inferences could be made concerning the impact of policy upon household decision-making and food security. Overall the study provides fertile ground for future work. It is hoped that by integrating PRA exercises and ABM it will be possible to create a collaborative framework that promotes interaction between scientists, policy makers and stakeholders, alike. Furthermore it has a potential use for the GCC countries: firstly, in adapting the ABM presented in the chapter alongside tools for PRA to analysis their own food production systems; secondly, in gaining a better understanding of the environments in other countries that could be a target for FDI in relation to food production.

The final **Chapter 11** by **Mayami Abdulla** focused on GCC initiatives towards utilising renewable solar energy and seeks to answer the question: is there a coherent and integrated Plan? The chapter puts this question into a clear context. The GCC countries receive an abundant amount of daily solar radiation. Solar radiation is an excellent free and non-diminishing alternative to fossil fuel for generating electrical power in countries, such as the GCC, that have sufficient amounts. In response to the escalating demands for energy, the GCC countries are experiencing an accelerating growth in the initiatives utilising renewable energy, primarily solar energy. These initiatives, in contrast to their conventional fossil fuel-based counterparts, align well with the goals of a green economy and sustainable development. Furthermore, they provide evidence that the GCC countries are meeting their commitments on the global stage to contribute to efforts to employ more environmentally friendly resources for electricity generation. However, despite the huge governmental attention being paid to employing solar energy in electricity production in the GCC countries, there is still a lack of a coherent, integrated and comprehensive plan for endeavours towards the exploitation of solar energy in the region. The chapter helps to fill this gap by shedding light on the impact policy measures can have on achieving the sustainable utilisation of solar energy. This is done by reviewing and analysing a number of sources of information including: academic literature, international organisations' policy and practice reports and press releases related to the solar energy field. The drivers that propel the GCC member states to embrace the renewable solar energy as an integral part of their energy portfolios are analysed. The chapter further identifies the potential remunerations that will be earned from utilisation of solar energy in the GCC countries and reports recent examples of the associated initiatives in these countries. The chapter concludes with **Abdulla** recommending tactics that will reinforce a coherent solar energy policy in the GCC countries, giving special emphasis to the policies leading to an extensive domestic diffusion of solar energy appliances within these countries. The tactics include educational programs to aid the GCC countries in confronting the envisaged skills shortages that are likely to arise in the growing indigenous solar energy job market.

References

ARSSCP (2009) Arab Regional Strategy for Sustainable Consumption and Production [Final Draft] [Accessed 12th Dec 2014: http://www.unep.fr/scp/marrakech/publications/pdf/Final%20Draft%20 Arab%20Strategy%20on%20SCP%20-%2006-10-09.pdf]

Elkington, J. (1997). Cannibals With Forks: The Triple Bottom Line of 21st Century Business. Capstone, Oxford.

UAE Ministry of Cabinet Affairs (2014) UAE Vision 2021, Ministry of Cabinet Affairs. [Accessed 12th Dec 2014: http://www.moca.gov.ae/?page_id=620&lang=en]

UNCED (1992). United Nations Conference on Earth & Development. Rio de Janeiro, 3rd-14th June.

UNEP (2014) Marrakech Process, United Nations Environment Programme. [Accessed 12th Dec 2014: http://www.unep.org/resourceefficiency/Home/Policy/SCPPolicies/MarrakechProcess/ tabid/55816/Default.aspx]

UNFCCC (1992). United Nations Framework Convention on Climate Change. New York.

1

Transitioning toward Sustainable Consumption and Production: Opportunities and Challenges in the Gulf Cooperation Council

Kassem El-Saddik

1. Introduction

The economic development of Gulf Cooperation Council (GCC) states over the last three decades has been achieved at unprecedented rates but at uncalculated cost to the region's scarce natural and environmental resources. Besides exporting oil and gas, GCC are increasingly consuming bigger share of their underground reserves to fuel their rapid economic and urbanization growth. Demand for electricity is soaring and will grow inevitably by seven to eight percent annually for the next decade to meet the escalating need for air conditioning and water- generated mostly from energy-intensive desalination processes. Balancing their socio-economic development, coupled with inefficient energy use and damaging environmental impact, becomes an inevitable choice that GCC states are facing. The challenge will more likely exacerbate in the face of climate change and the scarcity of water resources in the region. More recently, they have embarked on many initiatives toward sustainability in both the energy and water sectors, as affirmed in the Arab Regional Strategy for sustainable consumption and production (2009), as well as other national plans.

Based on the overall Sustainable Consumption and Production (SCP) framework adopted in the Marrakech Process and as stipulated in Agenda 21, Johannesburg Plan of action and Rio+20 Outcome Document "The Future We Want", this chapter attempts to draw the GCC-specific sustainable Consumption and Production framework, questions the key issues related to the region's SCP priorities, examines the progress induced since the Arab Regional Strategy for sustainable consumption and production 2009, explores opportunities and challenges facing GCC states, and suggests policy options to drive GCC states' transition toward sustainable consumption and production patterns.

2. Sustainable Consumption and Production Discourse

2.1. Definition and Evolution of the Concept

Sustaining the consumption and production patterns, in simple terms, aims at increasing the net welfare gains from economic activities by reducing resource use, environmental degradation and pollution, while enhancing the quality of life. It is therefore a powerful lever to accelerate the transition to achieve sustainable development. Barber (2012) has traced SCP discussion back to Schumacher's publication "*Small Is Beautiful*", noting that the addiction to fossil fuels and abuse of natural capital put modern civilization on a "collision course." To change course, Schumacher (1973) advised, "*We must thoroughly understand the problem and begin to see the possibility of evolving a new life-style, with new methods of production and new patterns of consumption*". Around the same time, the Club of Rome released its famous publication "The Limits to Growth", in which Meadows prescribed a shift away from the economic growth paradigm in order to avoid overshoot and collapse. Two decades later, 108 heads of state adopted the *Agenda 21* in 1992 acknowledging that "*the major cause of the continued deterioration of the global environment is the unsustainable pattern of consumption and production*". Among other things, these heads of state agreed in Principle 8 of the Rio Declaration that to "*achieve sustainable development and a higher quality of life for all people, States should reduce and eliminate unsustainable patterns of production and consumption and promote appropriate demographic policies*". One of the most notable propositions from Rio that becomes central to SCP discourse is the call for "*new systems of national accounts*" that do not depend on economic growth but rather on "*new concepts of wealth and prosperity which allow higher standards of living through changed lifestyles and are less dependent on the Earth's finite resources and more in harmony with the Earth's carrying capacity*" (UNEP, Switch Asia Policy Support, Sustainable Consumption and Production: A handbook for Policy Makers, first edition, 2012). Seven years later, in 1999, the UN General Assembly updated its Guidelines for Consumer Protection to include a section on sustainable consumption. Ten years later, in Johannesburg, SCP was declared one of the three "overarching objectives of, and essential requirements for sustainable development" besides environmental protection and poverty reduction. A 10-year framework programme was initiated to guide and accelerate the national and regional transitions towards sustainable consumption and production. It was designed to tackle and promote delinking economic growth and environmental degradation through improving efficiency and sustainability in the use of resources and production processes and reducing resource degradation, pollution and waste. The framework was then further refined in what was then called the Marrakech Process.

Box 1 UN Guidelines on Sustainable Consumption (UNEP, 2012)

In 1985, prior to the Rio conference in 1992, the UN General Assembly adopted the UN Guidelines for Consumer Protection, an international framework that provides support for activities of consumer organizations as well as guiding principles for the development of national consumer protection legislation (UN 2003b). The guidelines included the rights to: safety, information, choice, representation, education, redress, a healthy environment and basic needs. Later on, to reflect growing concerns in unsustainable patterns of consumption and production and the need to bolster government and other stakeholder efforts to promote sustainable consumption, the guidelines were expanded in 1999 with Section G on "Promotion of Sustainable Consumption". Recently, new developments in technology, forms of social organization, and business practices, present new challenges that are not currently reflected in the Guidelines. The United Nations Conference on Trade and Development (UNCTAD) has announced that it will start a new revision of the guidelines, to be tabled for adoption in 2014.

The guidelines call on Governments (in partnership with other stakeholders) to take leadership in several actions, including:

- "develop and implement strategies that promote sustainable consumption through a mix of policies"
- "removal of subsidies that promote unsustainable patterns of consumption and production"
- "encourage the design, development and use of products and services that are safe and energy and resource efficient, considering their full life-cycle impacts"
- "impartial environmental testing of products"
- "safely manage environmentally harmful uses of substances and encourage the development of environmentally sound alternatives for such uses"
- "develop indicators, methodologies and databases for measuring progress towards sustainable consumption at all levels"

2.2. Sustainable Consumption and Production: The 10-Year Framework

In 2012 in Rio de Janeiro, governments reiterated from Agenda 21 and the Johannesburg Plan of Implementation, and recognized *"the fundamental changes in the way societies consume and produce are indispensable for achieving global sustainable development"* (UNGA, 2012). Besides, some key governance and funding enablers were adopted to facilitate the implementation of the 10YFP, namely assigning UNEP as secretariat, insisting on its multi-stakeholder nature, establishing a trust fund to implement it in developing countries. Accordingly, SCP is perceived as *"a holistic approach to minimizing the negative environmental impacts from consumption and production systems while promoting quality of life for all"* (UNEP, 2012). The underlying principles of SCP revolve around considering the potential impacts from all life-cycle stages of production and consumption process (figure 1), while:

1. Improving the quality of life without increasing environmental degradation and without compromising the resource needs of future generations.
2. Decoupling economic growth from environmental degradation by:
   - Reducing material/energy intensity of current economic activities and reducing emissions and waste from extraction, production, consumption and disposal.
   - Promoting a shift of consumption patterns towards groups of goods and services with lower energy and material intensity without compromising quality of life.
3. Guarding against the re-bound effect, where efficiency gains are cancelled out by resulting increases in consumption (UNEP, 2012).

It is important to note that the SCP life-cycle management is based on precautionary and preventive approaches covering the whole value chain. While considering the various phases of a product life cycle, it highlights both supply and demand side sustainable and efficient management of resources, and promotes the use of efficient (fewer resources and generate less waste) and environmental-friendly but productive and economically viable processes. It encourages capturing and reusing and/or recycling valuable resources, thereby turning waste streams into value streams. It opens a wide spectrum of policy options to support sustainable consumption and production.

3. The Consumption and Production Framework: The "Rentier" State Implications

Located on the west coast of the Arabian Gulf, Bahrain, Kuwait, Qatar, Oman, United Arab Emirates and Kingdom of Saudi Arabia, established in 1981 a political and economic union called the Gulf Cooperation Council (GCC) officially referred to as the Cooperation Council for the Arab States of the Gulf. The GCC economy has become one of the fastest growing economies in the world mostly led and managed by the governments. Fueled by oil and gas, GCC has witness unprecedented social and economic transformation. A large proportion of the region's wealth is driven by the exploitation of oil and its associated industries, along with the savings accumulated over decades and a recent real-estate boom –one of the largest worldwide. The region is the world's chief supplier of oil and gas hosting around 40% of world's crude oil and 23% of its natural gas reserves (Krane,

2010). The total revenues have increased by more than 58% between 2006 and 2012, from US$311.7 billion to US $737.5 billion (Oil Review Middle East, 2013); while the GDP per capita almost doubled in 4 years between 2004 and 2006 reaching US$ 21,000, then increased by 30% to US $29,900 in 2011 (Gulf Cooperation Council, 2010). Besides the huge quantities exported, domestic demand for oil and gas is increasing to fuel the rapidly expanding transport, real estate and industrial development, transforming GCC states into energy consumers. Demand for electricity has doubled over the last decade and will grow, mostly driven by the domestic, commercial and industrial sectors. It is estimated that the built environment, hence urban expansion, accounts for more than 75% of total electricity demand in each GCC state (Bachellerie, 2012). Such development path has been manifested through:

1. The sole reliance on hydrocarbons to fuel their energy-intensive industries;
2. A world-record population growth -driven mainly by labor emigration- estimated at 42% of GCC population in 2010 (AFED, Survival Options, the Ecological footprints of Arab Countries, 2012)
3. Unprecedented urbanization.

The region's dependence on fossil fuels and the unwritten yet practiced social contract- that is based on its natural resource revenues- produces a pattern of "natural unsustainability", which Luomi describes it as a built-in feature of the "rentier" element of the contemporary political economies of the GCC States (Mari, 2012). It is determined by growing "energy wasteful" tendencies influenced by the demographic, living patterns and lifestyle of the society on one hand, energy inefficient consumables and products on the other, as well as cheap and subsidized energy supplies.

Al-Thani confirms the current unsustainable trends in the GCC communities in the forward note in his book "The Arab Spring and the Gulf States", when he compares the new and old generations, and appreciated the fact that "*they (the old generation) never wasted food or money*" (Al-Thani, 2012).

Box 2 "Rentier" Class

Since the discovery of oil, a 'rentier class' has developed which generates an income through the collection of 'rents', namely through sponsoring, taking commissions (...) 'rentier' class limits economic development as it undermines productivity and provides minimal incentives for risk taking (Al-Thani (2012), p. 6).

In order to better understand, figure 1 attempts to draw the link between the consumption and production patterns in GCC. Though the framework is based on the life-cycle approach, it highlights the key drivers for production and consumption, namely "economic prosperity", and emphasizes the enabling factors and external environment influencing them. It illustrates mostly on the production and consumption of both electricity and water.

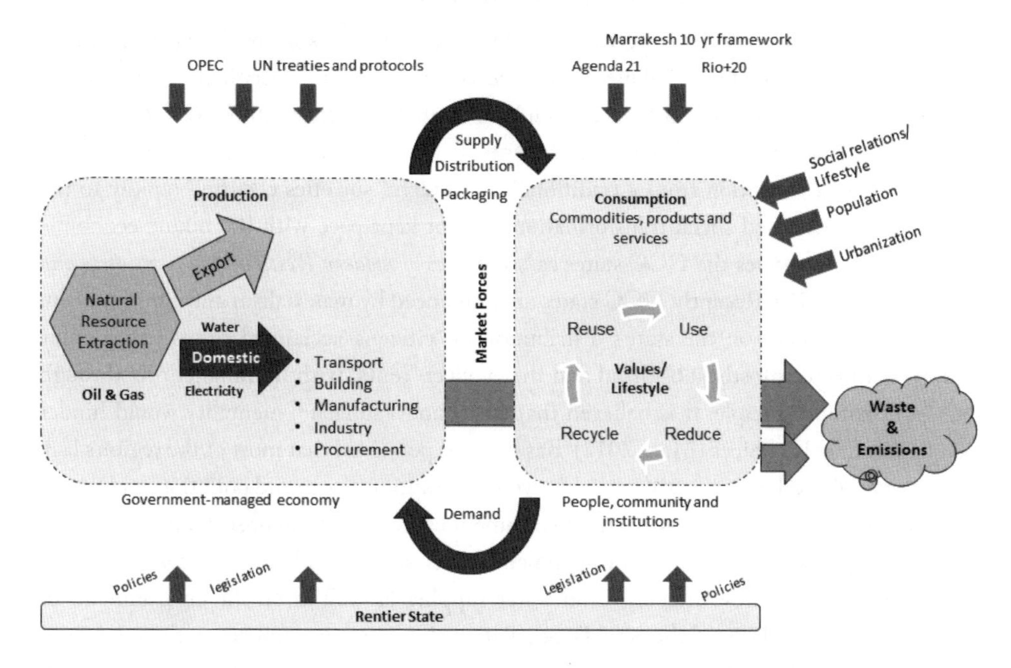

Figure 1 **GCC Consumption and Production Framework (Author)**

4. GCC Sustainable Consumption and Production Drivers

4.1. Population Growth and Urbanization

The GCC population has grown more than ten times during the last 50 years; from 4 million in 1950 to 46.5 million in 2010, with a skewed population pyramid exhibiting a male predominance at the productive ages of 20 to 45 years. Such distortion is driven by the high influx of foreign workers making around 42.7 percent of the total population in the region (AFED, Survival Options, the Ecological footprints of Arab Countries, 2012). The expat population in both wealthy states of Qatar and UEA soars to reach more than 80% of their population size. Driven by the economic prosperity, various population growth estimates for the year 2020 prevail. While the Economist Unit predicted the GCC population to grow by a 30% in the next decade, hitting 53.5 million by 2020 ((EIU), 2010), some local estimates done on the basis of the 2010 GCC census report, overshoots the EIU's and forecasted the region's population to exceed 60 million by 2020, mostly residing in urban cities. In fact, the urban population in the GCC states has grown from 40% in the 1980s to more than 90% of the total population at present, with the highest rates recorded in Kuwait (97%), Bahrain and Qatar (92%) (ESCWA, 2007). These growth scenarios raise serious flags related to the improving or maintaining of quality of life, meeting the escalating demand for energy, water and food, housing, transportation and waste management in urban settings.

4.2. Lifestyle and Social Consumption Patterns

From a human development perspective, four of the six GCC states are exhibiting a high HDI; two of them, namely Qatar and United Arab Emirates, ranked in the "very high" HDI category (HDR, 2012). A closer look at the HDI ranking suggests a correlation between the level of development and affluence and the amount of energy and water consumed and waste generated.

The GCC transition from a traditional to "modern" societies occurred rapidly to the extent that a sustained social transformation has not kept pace with the hiking economic one. Al-Thani describes the GCC states as "*exemplars of modern Western-style capitalism and consumerism*" (2012). Recently, GCC states are influenced by materialism and consumerism; the population relies on the state's distribution of "citizens' acclaimed rights"; they enjoy energy abundance mostly subsidized and they collect "rents" with minimal efforts through the sponsorship principle. It is believed that such "nouveau riche" mentality would hinder people sustainable choices (Mari, 2012). Besides, it is perceived that most of the region's lack infrastructural and legal incentives to curb overconsumption, waste and pollution, and whose population shows no sensitivity to the environment cause would run unsustainable lifestyle" (Al-Thani, 2012)). As such, GCC development seems to be gained at a high cost producing unprecedented pressures on energy and water supplies as well as on the environment. In fact, with less that 0.6% of the world's population, GCC states emit more than 2.4 % of global emissions (Raouf, 2008), and generate approximately 120 million tons of waste in 2009. A resident in Kuwait, Riyadh or Abu Dhabi is found to generate over 1.5 kg per day per capita, making it one of the highest levels worldwide. This consumer pattern is further aggravated when the ecological footprint of GCC countries - estimated at 239 million gha- outpaced its bio-capacity by seven times -estimated at 33 million gha. It is envisage that with the projected pace of development, the region's resources will be further stressed by the same conventional production and unsustainable consumptions schemes. Along with the expected impacts of climate change, the existing pressures are likely to intensify, hence put tremendous challenges on the survival options of the GCC States (AFED, 2012).

4.3. Energy Consumptiona

GCC rapid economic expansion has put tremendous strains on the region's natural resources, limits the export power (burning bigger quantities domestically consumes some of the oil allocated for export, hence abandon part of the state's revenues), and exposes the region globally given its carbon footprint. The demand for energy in GCC states has been maintaining a steep growth since 2005, and will keep it up to generate electricity, making the per capita energy consumption among the highest worldwide. The latter is expected to keep growing too, as suggested by the Economist Intelligence Unit (figures 2 and 3), to the extent that the average Qatari consumes energy nearly 10 times the global average and 53 times more than an average Yemeni. (AFED, 2008).

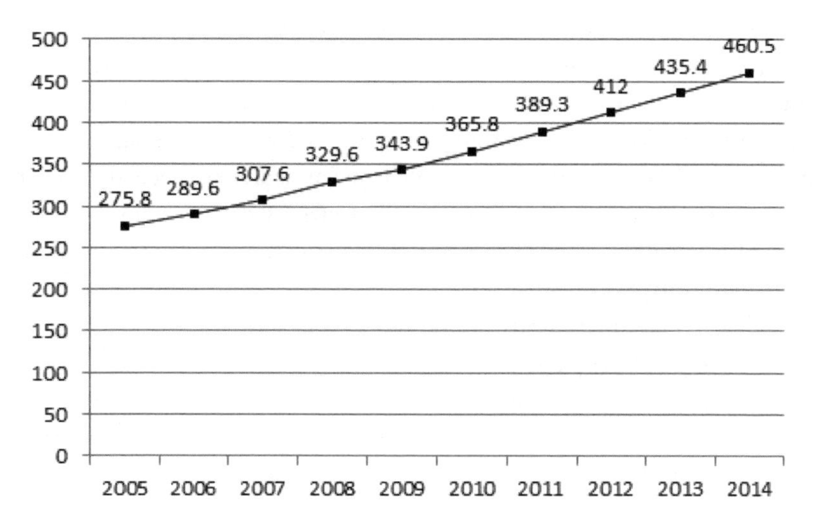

Figure 2 **Total Energy Use in GCC (million tonnes of oil equivalent, 2005-2014) (EIU)**

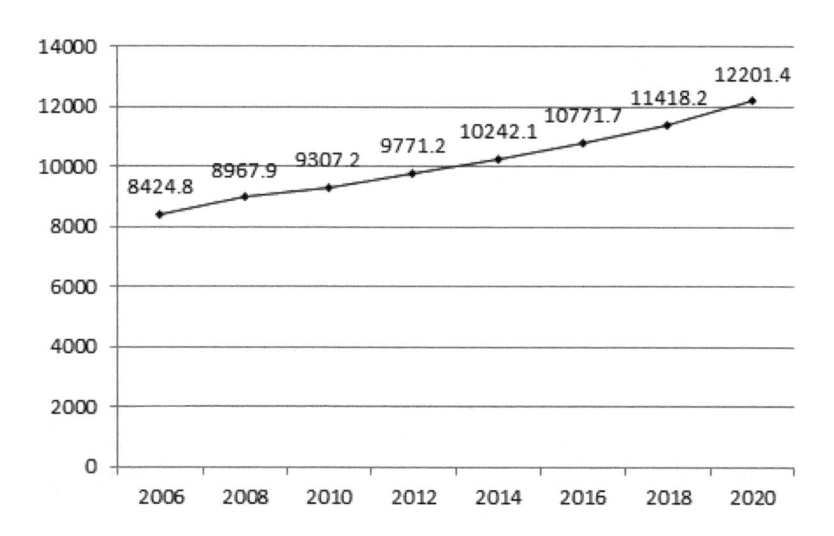

Figure 3 **Per capita energy consumption in GCC (2006-2020) (EIU)**

Overall the electricity generation capacity (measured in megawatts) has kept pace with the peaking demand, except for Kuwait that suffered electricity shortage and blackout in 2009. Qatar, on the other hand, after its 2008 over-demand, started witnessing a surplus. The latter has been exported through the GCC interconnection electricity grid. Domestic consumption constitutes more than half the electricity consumed annually. In 2009, it recorded 55% in Oman, 56% in Bahrain and 53% in Saudi, whereas, the commercial sector ranked the second consuming 28% in Bahrain and 18% in Saudi. Given the harsh climate and due to residing in energy-inefficient building, air conditioning alone is believed to account for more than 80% of the household electricity consumption (Bachellerie, 2012). To curb it down, rethinking electricity tariff and introducing energy conservation and efficiency measures are needed. It is estimated that the latter would save the building annual electricity consumption by a third. The savings could reach more than 40 percent if photovoltaic panels are installed on the roofs (Bachellerie, 2012).

The generated electricity feeds the various sectors and powers the energy intensive water desalination plants. The carbon footprint, reflected by the carbon dioxide emissions, associated with electricity generation clearly reflect that electricity and heat production account for more than 35% of the emitted carbon dioxide in GCC, except for Qatar whose CO_2 is mostly driven by other energy and gas intensive industries.

This is in line with the global trends in electricity generation that accounts currently for approximately 50% of global carbon emissions, given its high dependence on fossil fuel (70%)and the inefficiency which wastes around 65% of energy contained in fuel (El Khoury, 2012). A closer look at the table (1) clearly confirms the observations above and highlights anemerging contribution driven by the transportation sector, ranking third or fourth among other sectors.

4.4. Water Consumption

The Middle East and North Africa Region (MENA) and particularly the Gulf Cooperation Council have been classified as the most water scarce region in the world. While the world

***Table 1:* Carbon Dioxide Emissions from Fuel Combustion in GCC by sector (2009), (Bachellerie, 2012)**

GCC State	Electricity & Heat Production	Transport	Residential	Manufacturing & Industries	Other Energy Industries	Other Sectors
Kuwait	57%	14%	1%	15%	13%	0%
Oman	39%	14%	4%	21%	19%	3%
Saudi Arabia	39%	25%	1%	19%	16%	0%
UAE	39%	17%	0%	43%	1%	0%
Bahrain	35%	15%	1%	30%	19%	0%
Qatar	22%	11%	0%	33%	34%	0%

estimated average water availability per person is 7,000m^3/person/year; the region's average is around 1,200 m3/person/year thus putting more than half of MENA's population under conditions of water stress. A recent estimate suggests that Arab states in West Asia (Bahrain, Egypt, Iraq, Jordan, Kuwait, Lebanon, Oman, Palestine, Qatar, Saudi Arabia, Sudan, Syria, the United Arab Emirates and Yemen) will enjoy no more than 500 m^3 of fresh water per capita by the year 2025 (Worldbank, 2009). Obviously the situation differs from state to state due to different sources of water supplies associated with their geographical location and topological characteristics. In fact, the GCC states suffer the most due to their arid topography and lack of fresh water resources, except for Oman whose renewable water resource amount at 503 m^3 of fresh water per capita per year. GCC states are challenged with a double burden: they have very little renewable water resources but their consumption is the highest worldwide, hence invest heavily and costly in energy-intensive desalination infrastructure. The latter accounts for 57% of the world total desalination capacity. Saudi Arabia for instance is the world's largest producer; it produces 24 million m^3 of water per day from its 30 operating plants (Markaz, 2012). Such investments (estimated at 19.5 billion US Dollars for new infrastructure over the next decades (MEED, 2010)), however, are seldom coupled with other policy tools to rationalize unsustainable water consumption that is heavily subsidized. In fact, consumers pay on average no more than 60% of the production and supply cost, estimated at 2 USD/m^3. The figure reaches less than 2% in Saudi Arabia, 4% in Bahrain and 30% in Kuwait. In terms of usage, the agricultural sector consumes almost 70% of the regional water, but contributes to less than 1% of its GDP (Markaz, 2012). Municipal consumption stands the second on the list particularly in Bahrain, Kuwait and Qatar.

As per capita water supplies get tighter and investments exceed the nine-digit figures, water management needs to be strategically addressed by GCC governments to promote and optimize water use efficiency in the demanding agricultural, industrial, commercial and domestic sectors. In terms of inefficiency, it is estimated that non-revenue water (NEW) in Bahrain ranges between 20% and 30% of the produced water due to leakage, meters inaccuracy and unbilled consumption directed to firefighting and Mosques.

***Table 2* Water Consumption by sector in GCC (%)**

GCC State	Agriculture	Municipal	Industrial
Bahrain	44.5	49.8	5.7
Kuwait	53.9	43.9	2.3
Oman	88.4	10.1	1.4
Qatar	59	39.2	1.8
Saudi Arabia	88	9	3
UAE	82.8	15.4	1.7

While suggesting that water policies in the Arab region require improved management on both the supply and demand sides, the 2007 AFED report highlights the striking conflict in policies set at the highest levels in some of the GCC states, calling for the construction of water-intensive golf courses despite the high cost of desalination technologies (AFED, 2008). Besides, though governments are recognizing the current unsustainable consumption patterns, their attempts to curb domestic demand are still shy and not conducive. Setting tariff slab on water consumption is believed to be a key water management policy tool. However, it proves not to be practical except in societies with predominant expat population. In fact, access to low-cost water is perceived as a right to GCC national population. Kuwait did not proceed with introducing the tariff slab. It was highly politicized and no decision has been taken. Dubai however has succeeded in introducing it to incentivize consumers and rationalize consumption. However, the tariff slab was set on expatriates' consumption only. It excluded the nationals given its sensitivity (being a "right"). Obviously it did not curb down consumption, a slowdown in demand was observed (from 12.4% between 2005 and 2006 to a demand of 2.6% for 2008-2009, associated with an increase of registered DEWA customers from 10.8% to 19.5 % respectively) (MEED, 2010). Challenged with the sensitivity of introducing a tariff reform, states have reverted to other forms of policies to curb and control demand for water mainly focused on addressing inefficiency through introducing the water metering, cutting NRW and promoting public awareness.

5. Managing the Transition toward SCP: Opportunities and Challenges

5.1. Pre 2009

The GCC states have recognized the challenges they are facing with regard to their economic reliance on fossil fuel as the single resource driving their development. In 1998, the GCC Supreme Council convened in Abu Dhabi and ratified its "long-term Comprehensive Development Strategy for GCC States" –a 25 year strategic plan (GCC, 2011). Almost 10 years later, the same Council ratified a revised plan in which the GCC states acknowledged that the continuing dominance of a single resource (oil), in addition to climate change, scarcity of water, the "fatherly care" of the states' citizens, the population growth and imbalance and the increasing consumption powers are the real challenges they need to address individually and collectively. The Council then endorsed a series of strategic initiatives, within the framework of sustainable development, aiming at optimizing resource utilization, preserving the environment, intensifying investment in non-petroleum manufacturing, adopting modern water resource management to ensure sufficiency and efficiency, providing adequate resources for alternative energy including renewables and nuclear energy for peaceful purposes, and integrating the environmental dimension in policy making and strategy planning. Although at that time the strategy did not explicitly refer to "sustainable consumption and production", it took an SCP-featured approach towards sustainability and claimed it will be integrated into economic and social policies.

5.2. Post 2009: The Arab Strategy and Beyond

Many regional and international determinants have set the scene for the GCC to initiate its SCP strategy. The international momentum triggered by Marrakech process and the preparation work for Rio+20 have reinforced the concept and reaffirmed a global consensus toward putting the 10 year framework for SCP on track. Besides, the escalating unsustainable pressures on the hydrocarbon resources, mainly domestically, coupled with the unprecedented and unforeseen electricity blackouts in some GCC states (the Kuwait blackout in 2009), energy inefficiency and wastage (that might reach two-third of the generated energy) (Bachellerie, 2012), and the carbon emissions associated with the domestic use (electricity generation accounted for more than 40% of the region's carbon dioxide emissions in 2009) played key roles in shaping the SCP discourse in the region. In response to that, the Council of the Arab Ministers Responsible for Environment (CAMRE) commissioned the Joint Committee on Environment and Development in the Arab Region (JCEDAR) to draft the first Arab Strategy for Sustainable Consumption and Production in 2009. The Arab SCP strategy aims at "*promoting the concept of sustainable consumption and production in the Arab region by encouraging the utilization of products and services that ensure environmental protection, conserve water and energy as well as other natural resources, while contributing to poverty eradication and sustainable lifestyles*" (JCEDAR, 2009). It identifies six priority areas, namely the energy for Sustainable Development and water resources management (being top on the priority list), followed by key 'priority implementation areas' such as Waste Management; Rural Development and Eradication of Poverty; Education and Sustainable Lifestyles and Sustainable Tourism. At the sub-regional and national levels, SCP has been addressed in various sectoral development or environmental plans and policies. Most recently, SCP has been integrated to a large extent through the national Green Economy agendas and initiatives many GCC states have embarked upon.

Various policy instruments have been put forward in the region as tools to promote SCP in the energy sector. Most of them are however more focused on supply-side rather than on demand-side measures. It is argued that the demand-side management through the conventional tools would disturb the established social contract within the states. Unfortunately the region has limited incentives imposed on unsustainable consumption, and the information on the demand side of SCP remains scarce. Through the Arab Strategy, the governments –among which the GCC- expressed serious needs to study and analyze the sustainable lifestyles, with youth and women as priority targets. Given the links between energy and water, GCC need to pursue sustainable production and consumption in an integrated life-cycle approach addressing the two pillars of consumption and production along two timeframes, an immediate/ short-term and a long-term one; central to the transition lies a up-to-date legislative enabler.

5.3. The Opportunities and Challenges

Managing domestic demand for both fossil fuels and electricity remains a key challenge in a region that provides its abundant supplies at a subsidized rate as a form of resource redistribution under the "rentier" state principle. Besides the political commitment that seems to be currently non-negotiable, GCC citizens consider they have the right to consume the state's resource. Unfortunately, this has been done in a wasteful manner with little regard to its finite nature, and hence with little interest in resource conservation and efficiency. Should the resource be of an infinite renewable nature, it is argued that the political sensitivity will be minimal and the potential to influence the citizens' consumption patterns be more promising. In fact, their rights to consume will be respected and ensured while contributing less to the environmental degradation.

Therefore, on the long term, GCC states should craft means to reduce energy consumption while accelerating the transition toward alternative energy supplies to maintain their citizens' rights in the natural resources. Shifting into renewable alternatives seems to be the ultimate scenario. Renewable sources such as the solar and wind energies are abundant, and GCC states have taken serious pilot steps in that path. In fact, some policies and programs have been recently launched. The Kingdom of Saudi Arabia has embarked onto its ambitious plan to install 41 gigawatts (GW) of solar energy by 2032, with 25 GW of power generated using concentrated solar power (CSP), while photovoltaic technology supplies the balance. Unfortunately, the Kingdom has not issued any new regulation to influence energy consumption behavior; neither did it promote small scale localized renewable schemes, despite their direct impact on minimizing energy usage. It would be worth noting that crafting a regulatory framework to incentivize and promote the feed-in tariffs seems a promising opportunity yet to be seized.

Besides addressing the supply side, Saudi Arabia decided to create a serious market need for energy efficiency measures through the National Energy Efficiency Programme, the Saudi Green Building Council (SGBC) and the King Abdullah City for Atomic & Renewable Energy. Similarly, Abu Dhabi has invested around $600 million in solar energy to construct the 100 MW Shams-1 CSP plant to create other alternative supply. In addition to the pilot zero-emission MASDAR City, UAE has put serious efforts in the short term to establish a need for resource efficient housing through the Green Building policy among others. On another note, the private sector has been playing a key role in implementing some voluntary smart but small-scaled energy-efficiency changes, such as hotels that supply a key card to switch on lighting, and solar panels, etc... Demonstrating the private sector commitment to invest in efficiency and alternative energy projects should be leveraged once the right policy incentives are in place. While predicting that GCC states will follow global best practices to increase the energy efficiency of buildings, including introducing (voluntary) sustainable building codes, Mari Luomi questioned its implementation in the absence of the policy incentives (Mari, 2012). Therefore the proper legislative and policy tools need to be enacted to bring down the unsustainable consumption patterns. Besides,

Governments in the region are the main driver of economy given their hugr investments in the various sectors. Government procurement seems then a promising stream to shift toward green economy and sustainability, as advocated by all GCC states. The public sector in GCC, consisting of the various government agencies and semi-governmental companies, consumes hugely on goods, services and equipment. Procuring energy-efficient equipment will not only save the governments' budgets and energy, but would influence the people and the private sector's attitudes too. It will create demand for green goods and services, and drive the market to supply more. A local demand for "green" (energy-saving) machinery and equipment will then impact positively the consumption; however, serious considerations should be taken not to risk local industries and set indirect barriers in favor of the big manufacturers and suppliers.

Furthermore, a slab energy tariff scheme could be a suitable option to manage domestic over-consumption by raising tariffs on consumers with higher-than-normal electricity consumption. The same might apply on the commercial and industrial sector while considering incentives for small and medium enterprises.

Similarly, the GCC States' emphasis on water supply expansion is always their preferred option to avoid adopting politically sensitive recommendations that might shake the adopted "social contract" format. Hence, most of the measures introduced to address the root causes of unsustainable consumption have been limited. The introduction of basic utility charges on electricity in Qatar and water in Dubai in 2008 was an initial tentative move toward enhancing awareness of the cost of consumption and the need for efficiency. However both initiatives, while targeting expatiates only, excluded the communities that exhibit a high standard of living and wasteful patterns of consumption.

The direction of the Gulf States' initiatives to boost capacity rather than change habits of unsustainable consumption may defeat the whole purpose. In fact, an increase in supply will be caught up with an increase in demand, creating a "rebounding effect". This is clearly illustrated with the growing demand for water in Oman by 25% triggered by the commissioning of both the Sur and Barka 2 desalination projects in Oman in 2009 (MEED, 2010). Rolling out subsidies to shift the patterns of wasteful consumptions toward more sustainable ones and injecting supplies will project the wasteful consumption of water to high levels. Immediate short and medium terms options need to be explored and implemented. On the demand side, governments shall attempt to curb consumption by:

1- Assessing the best tariff reform that does not jeopardize the social contract. A slab tariffs is still a suitable option.
2- Managing water supplies inefficiency through cutting non-revenue water (NRW) by investing in water connection and enforcing the use of recycled water. In fact, creating a need for and incentivizing the use of treated wastewater would be exemplar. In fact,

70% of fresh and desalinated water can be saved and diverted for domestic use once the agriculture, landscaping and commercial sector shifts into using tertiary treated wastewater.

3- Attracting the appropriate format of public-private partnership (PPP) to manage the water supply, without jeopardizing the State's ownership of its resources, will set the proper pricing scale for water usage and influence consumers' behavior. It would relieve the governments from some financial burden by freeing huge amount of oil and gas to be exported.

6. Conclusion: The Way Forward

The path is becoming clearer for the GCC transition toward sustainable consumption and production. Besides the above mentioned short and medium term options, there is no alternative but to invest in research and development as well as in the infrastructure of renewable energy. The main new factors encouraging the use of renewable energy can be summarized by the GCC states' genuine desire to minimize burning of oil to generate electricity and water on one hand, the concerns over the lost export earnings associated with the increased burning hydrocarbon domestically, the need to improve both electricity and water production and distribution efficiencies, as well as to conform to global efforts to cut carbon emissions.

GCC states should seize the opportunity of its wealth abundance and the rising oil revenues to invest in alternative energy technology and harvest the infinite solar energy that is available all year long in the region. In fact, the late development in the renewable energy has reduced the financial barriers for adopting alternative energy sources, which were once unfeasible. GCC states have recognized so and set ambitious plans to move forward sustainably, as manifested in the Arab SCP strategy. Solar energy is increasingly becoming the attractive option. In fact, Abu Dhabi plans to have renewable energy capacity equivalent to 7 per cent of annual demand in 2020, or 1,500MW. Dubai has also announced plans to build solar capacity. Regulators in Oman, meanwhile, are studying the country's first commercial solar power project proposal. Kuwait too has announced plans to procure solar power capacity (MEED, 2012). However, the large scale infrastructure of renewables needs to be coupled with localized schemes to generate electricity that feed surplus into the main grid. The key for renewables to flourish is for the GCC governments to develop and enact the policy and legislative enabler to support such transition.

References

AFED. (2008). *Arab Environment Future Challenges.* AFED.

AFED. (2012). *Survival Options, the Ecological footprints of Arab Countries.* Arab Forum for Environment and Development.

Al-Thani, M. (2012). *The Arab Spring and the Gulf States.* Profile Books.
Bachellerie, I. (2012). *Renewable Energy in the GCC Countries: Resources, Potential and Prospects.* Gulf Research Center.
Barber, J. (2012). *Mapping communities of practice toward SCP.* Integrative Strategies Forum.
(EIU), T. E. (2010). Retrieved March 18, 2013, from The GCC in 2020: Resources for the future: www.commodities-now.com/component/attachments/download/142.html
El Khoury, G. (2012). *Carbon Footprint of Electricity in the Middle East.* Retrieved March 19, 2013, from www.carboun.com/energy/carbon-footprint-of-electricity-in-the-middle-east/#more-3585
ESCWA, U. (2007). *Compendium of Environment Statistics in the ESCWA Region.* UN.
GCC, E. (2011). *The Revised Long –Term Comprehensive Development Strategy For the GCC States.* Retrieved April 15, 2013, from http://sites.gcc-sg.org/DLibrary/download.php?B=469
Gulf Cooperation Council. (2010, April). Retrieved April 19, 2013, from www.gccsg.org/ statistics
HDR. (2012). *Human Development Report.* UNDP.
JCEDAR. (2009). *The first Arab Strategy for Sustainable Consumption and production* . LAS, CAMRE.
Krane, J. (2010). *DSG policy Brief no. 8: Energy Conservation Options for GCC Governments.* Dubai: DSG policy Brief no. 8.
Mari, L. (2012). *The Gulf Monarchies and Climate Change.* Georges Town University CIRS, School of Foreign Affairs in Qatar.
Markaz. (2012). *GCC Water Sector: Markaz Infrastructure Reports* . Kuwait Financial Center.
MEED. (2010). *Gulf looks to curb water usage.* Retrieved March 19, 2012, from www.meed.com/sectors/water/water-supply/gulf-looks-to-curb-water-usage/3078418.article?sm=3078418
MEED. (2012). Retrieved March 2013, 2013, from www.meed.com/supplements/2012/middle-east-electricity-2012/economics-turn-in-favour-of-solar-power/3127445.article?sm=3127445
Oil Review Middle East. (2013, March 19). Retrieved April 20, 2013, from GCC's oil and gas annual earnings hit US$737.5 billion: www.oilreview.me/industry/gcc-s-oil-and-gas-earnings-hit-us-737-5-billion
Raouf, M. A. (2008). *Climate Change Threats, Opportunities and the GCC Countries.* Middel East Institute, Policy Brief 12.
Shumacher, E. (1973). *Small Is Beautiful: Economics As If People Mattered, Small is Beautiful.* Blond & Briggs.
UNEP. (2012). *Global Outlook on Sustainable Consumption and Production Policies.* UNEP.
UNEP. (2012). *Switch Asia Policy Support, Sustainable Consumption and Production: A handbook for Policy Makers, first edition.*
UNGA. (2012). Resolution 66/288 "The Future We Want", A/RES/66/288. *Rio+20.* United Nations.
Worldbank. (2009). *Sector Brief "Water Resource management in MENA".* Retrieved March 19, 2012, from www.worldbank.org/MENA

2

In Search of Sustainable Development: Economic, Environmental and Social Achievements and Challenges Facing the United Arab Emirates

Mhamed Biygautane and Justin Dargin[1]

1. Introduction

The past two decades have witnessed a shift in the strategies of governments in the Gulf Cooperation Council (GCC) from the establishment of sound infrastructure projects and state building into the exploitation of natural resources to achieve sustainable development. This is reflected in the economic, environmental, and social aspects of policymaking in the region. This chapter addresses the issue of sustainable development (SD) in the context of the UAE, and demonstrates the relentless efforts the government has made to "sustain" its economic, environmental, and social growth. It argues that while the UAE has made noticeable progress in achieving its sustainability objectives, and is in the right direction towards its 2021 vision, it still faces numerous mounting challenges before it can boast the success of its SD initiatives.

One of the main challenges that face academics, scholars and practitioners studying SD is to provide a specific definition of this term. Currently there is no consensus on its definition as it consists of numerous interrelated fields like economic, social, and environmental sciences" (Ciegis, et al. 2009). The term first appeared in the World Commission on Environment and Development (WCED) 1987 report that describes SD as an approach that "meets the needs of the present without compromising the ability of future generations to meet their own needs" (1987). The Brundtland report (1987) argued that SD should meet "the needs of the present generations without compromising the ability of the future generations to meet their own needs, in the fields of environmental protection, economic growth and social equity" (as cited in ICTI, p. 78). The UN Commission on Sustainable Development (UNCSD) provides a relatively similar definition of SD and stresses the significance of economic and social aspects of human life by stating that: "sustainable development must overcome its environmental degradation, but it must do so without forgoing the needs of economic development, social equality and justice." According to the definition given by neoclassical economists, SD is the "efficient use of natural resources," in a manner that reduces their negative impact on the environment (Rennings & Wiggering, 1997). The social dimension of sustainability; however, focuses more on growth rather than development which is the case with the economic approach towards sustainability (Lele 1991). Some of the sustainability issues that need to be addressed involve how development often affects not only the environment, but also society and the economy, and the three elements are closely connected. Hence, governments are often faced by a dilemma between fostering economic growth on the cost of the environment, or vice-verse. Some of these issues involve urban regeneration, community development, energy and transport, ecology and pollution, climate change, oil and other resource efficiencies, carbon footprints, and corporate responsibility.

The UAE has constantly demonstrated its commitment to raise environmental awareness of sustainability and "implement innovative solutions to protect and sustain the environment using new, energy-efficient technologies to reduce its carbon footprint" (UAE's Vision 2021). The government's motives are directed toward making both Emirati citizens and expatriate residents more responsible and aware of the nation's "ecological deficit" by promoting environmental awareness, mainly through "preventive measures such as reducing carbon dioxide emissions, and regulation to defend fragile ecosystems from urban development" (UAE National Charter, 2011).

The chapter is organized as follows. The second section assesses the sustainability of the UAE's economic growth and the government strategies to diversify its economy to guarantee its growth in the short and long runs. The third section analyzes the environmental achievements of the UAE over the past forty years, and the UAE's government initiatives to curb Cos emissions. The fourth section addresses the social development indicators and performance of the UAE's government in creating social programs to enhance the quality of life for its citizens and residents. The fifth section is devoted to the challenges that the UAE still faces in its path toward achieving more sustainable development at the economic, energy, environmental, social and demographic, and cultural levels. The conclusion and recommendations section introduce the way forward that the government of UAE should consider in order to effectively manage its sustainability challenges.

2. Economic Development and Performance

The period from 1971 to 2014 witnessed fundamental changes in the UAE's economic development. That encompasses the monetary, fiscal, foreign trade and developmental strategies. These positive changes resulted from numerous internal and external factors that affected the country's direction and created both opportunities and challenges. There are two major turning points that have contributed to the economic miracle in the history of the UAE. The first one was the creation of a united federation in 1971 with a robust political structure that led to the use of one currency in the country (Davidson 2005). The other was the discovery and production of oil, the revenues of which have been effectively utilized to build state-of-the-art infrastructure, and diversify the economic activities and resources of the UAE (Peck 2007).

The year 2012 is considered a momentous year in the economic history of the UAE. It was the year when the real GDP expanded by around 4.4 percent to reach AED. 1.02 trillion. "It was the first time that GDP surpassed the AED 1 trillion mark" according to Rashid Al Suwaidi, director general of the UAE National Statistics Bureau (UAE Interact, 2013). Al Suwaidi further commented that "there was positive growth in all non-oil sectors last year… this was the main factor in growth in the overall GDP and this has allowed the UAE to maintain its position as one of the largest economies in the world" (UAE Interact, 2013). Moreover, it was the year that the UAE was ranked among the top five countries globally for quality of macro-economic environment and infrastructure (Global Competitiveness Report, 2013).

Figure 1 **GDP (Current US$) of the UAE**

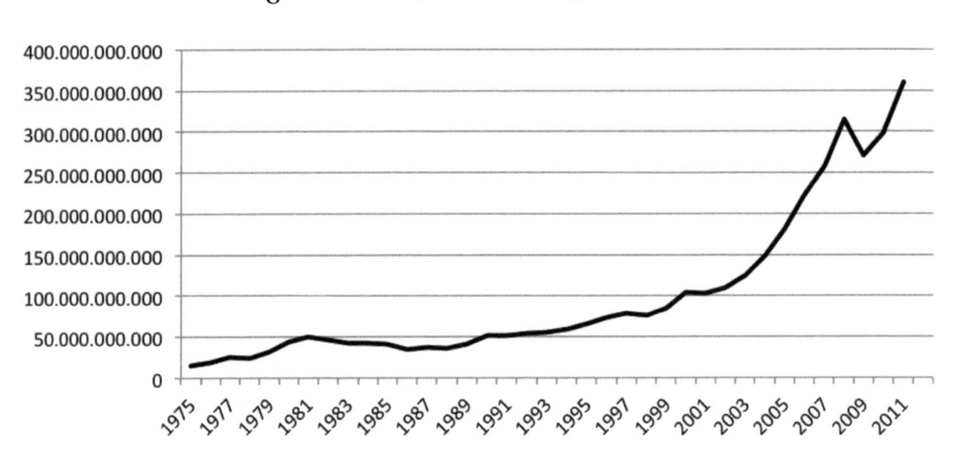

Source: World Bank Data Bank (2013)

As Figure 1 demonstrates, the UAE's GDP has grown exponentially since 1971. In 1990, the UAE's economy exceeded the $50 billion ceiling under which it had operated for the previous two decades. Only ten years later, the UAE's economy made a phenomenal jump from $100 billion in the year 2000 to reach $300 billion in 2008. The UAE's GDP growth has been constant, robust, and phenomenal. The GDP witnessed growth of almost 50 percent from the year 2006, when it was $175 billion, to $270 billion in 2010 (World Bank 2013). This makes the UAE the second largest economy in the Arab world after Saudi Arabia. Moreover, Al Mansouri, the UAE's Minister of Economy, stated that the UAE's GDP has grown by more than 200 times since 1971 to reach $360 billion in 2012 (UAE Interact, 2013).

The development of the UAE's economy in the pre- and post-1971 periods is an unprecedented success story in the region. Before the discovery of oil, the UAE's major source of income was from pearling and fishing. However, the period from 1972 to 1998 witnessed GDP growth of 264 percent to reach US$1,471 million in 1998 from the mere $46,340 million it had in 1972. Fifty years ago, when the grandparents of today's youth were starting their families, the country had no central electrical grid, indoor plumbing, telephone system, public hospital, or modern schools. Men washed their kanduras (a long traditional gown) in the sea and dried them in the sun. The seven Emirates that are part and parcel of the UAE's federal system today were basically villages composed of huts and unpaved streets (Walters et al. 2006).

The UAE's economic development witnessed substantial growth since the year 2000. This was primarily due to the rising prices of oil and attraction of foreign direct investment. The gross domestic product (GDP) grew by 12 percent in 2002 and remained above 8 percent until 2006. This created a booming investment climate and placed the UAE among the fastest growing economies in the world. Figure 2 below shows the percentage fluctuations in the UAE's GDP from 1976 to 2010. Many international events and economic and oil crises affected the growth of the GDP. However, the resilience of the economy and the strength of its foundations made it sustainable and strong. The expected GDP growth in 2013 for non-oil domestic products is 4.3 percent, compared to 3.8 percent in 2012 (IMF, 2013). In 2015, the IMF is predicting the economic growth to touch 4.5 percent signaling quite clearly that the local economy is marking a prudent recovery from the global financial crisis that hit the world in 2007 (IMF, 2014).

Figure 2 **UAE's GDP Growth Percentage**

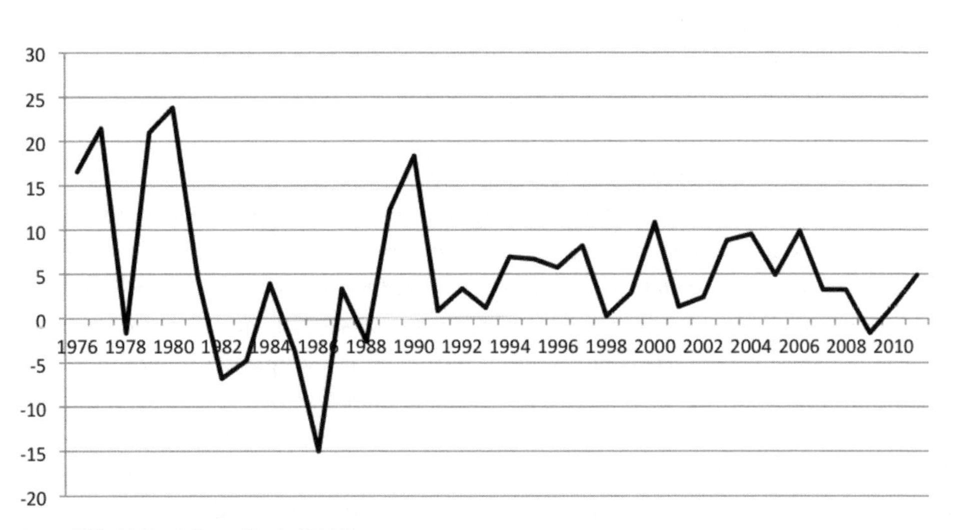

Source: World Bank Data Bank (2013)

3. Environmental Achievements

The massive economic and population growth that the UAE has witnessed since 1971, have put a significant amount of pressure on its environmental sustainability. Taking the environmental achievements of the UAE since its union into consideration is essential to assess its successes and shed light on the challenges that still face it.

The environmental threats that the UAE faces are numerous, and similar to those faced by other countries globally. These include industrial pollution, air pollution, overhunting of birds and fish, and mismanagement of natural resources. Overgrazing has been a direct cause of the desertification of many areas in the country. The UAE has developed various strategies and consulted with numerous consulting firms nationally and internationally to develop studies and conduct comprehensive assessment within the country to identify opportunities and handle threats to its environmental sustainability (Vidican et al. 2010, Mondal et al. 2014).

The government of the UAE is aware of the threats facing its environment and natural resources. Therefore, it has prioritized their preservation and protection, not only through governmental legislation and laws, but also through collaboration with numerous non-governmental organizations whose mission is to study and evaluate the health of habitats in the country (Sgouris et al. 2013). Federal laws have been put in place to unify the efforts of the UAE as a whole to safeguard its natural resources. Moreover, each Emirate also has its own strategic objectives that are aligned with the federal ones. Examples of significant legislation are Federal Law No. 7 of 1993[2], which established a Federal Environment Agency that created a coherent and clear framework to regulate and safeguard against environmental issues in the country. Another example is Federal Law No. 24 of 1999[3], which stipulates protection and development of the environment. This law has been influential in guiding the activities and focus of other Emirates to establish organizations and legal frameworks to protect the environment.

UAE's Government Initiatives to Reduce CO_2 Emissions

The UAE is aware of the challenges resulting from the continuous emission of CO_2 gas, and has embarked on numerous governmental and non-governmental initiatives to curb its impact on the atmosphere and its ecosystem in general. One major aspect in which the UAE has excelled and outpaced other countries in the world is its use of natural gas for electricity production. Since natural gas is proven to be less polluting compared to other sources of energy, the UAE's strategy is to utilize less coal and more natural gas for the production of electricity. The UAE produces 98 percent of its electricity from natural gas, while other Middle Eastern countries produce around 65 percent of electricity from natural gas. On the other hand, OECD countries produce only 20 percent of electricity by using natural gas (World Bank 2013).

Figure 3 shows the exponential rise of total electricity net consumption in the UAE since 1980s. For example, from 1990s to 2010, the total electricity consumption increased four times. This is an ultimate result of two factors: 1) the massive industrial and developmental trajectory the UAE has embarked on, 2) fast population growth that requires additional electricity support. According to DEWA's statistics (2012)[4], 39% of electricity consumption is residential, 28% commercial and 12% industrial. Furthermore, the UAE's energy demand is increasing by 9% every year which is three times the world average (The National 2013).The UAE is diversifying its economy to industries that are energy-intensive. Petrochemical industry is a field that the UAE tries to empower for the next twenty years. This means both focusing on oil refinery and production of petrochemical products, and also the use of petrochemicals to produce other materials like cement, glass, iron and steel... etc.

From 1970 onward, the population growth has been exponential in Dubai and the UAE. The world's population growth has been relatively stable, while in Dubai it picked up substantially. Expatriates from all over the world were attracted to the UAE and Dubai's economic model and wanted to take part of it. In 1900 about a quarter of Dubai's tiny population were foreigners, mainly from South Asia. Dubai's population was less than 40,000 people in 1940, and actually fell through 1954. By 1968, when oil was discovered, its population was roughly 40% foreigners. In the 35 years from 1975 to 2010 Dubai's population skyrocketed by ten-fold, to overwhelmingly 2 million foreigners (Estimates vary; the Dubai Statistics Centre put it at 1.9 million in late-2010, dsc.gov.ae). This growth was 60 times that of the world population's growth -- it is unprecedented, and enabled by a number of factors unique to the UAE including a wealthy federal system, an enlightened monarchy that has maintained security, safety and stability in the country.

The UAE has enacted a number of legislation, laws and strategies to ensure abidance by international standards in reducing CO2 emissions. In 1996, the UAE's government signed the United Nations Framework Convention on Climate Change, in which it recognized the urgent need to tackle climate change challenges, especially the emission of CO_2 gas (UAE's Ministry of Energy 2006). Therefore, it carried out major policies aimed to align economic growth with environmental sustainability. In 2008 for example, the Environmental Agency - Abu Dhabi (EAD) launched its Environment Strategy 2008-2012, in which it specified regulatory frameworks related to tourism, construction, emissions of CO_2 gas, and other sectors that are critical for achieving the targets of the UAE in lowering gas emissions (Environment Agency Abu Dhabi 2012). Moreover, the Federation pays closer attention to greenhouse gas emissions from different industries, conducts periodic studies and assessments of the impact of CO_2 gas emissions on the UAE's environment, and encourages the gradual elimination of these gases and utilization of other natural sources (Benkari 2013).

Figure 3 **Total Electricity Net Consumption in the UAE (Billion Kilowatt hours)**

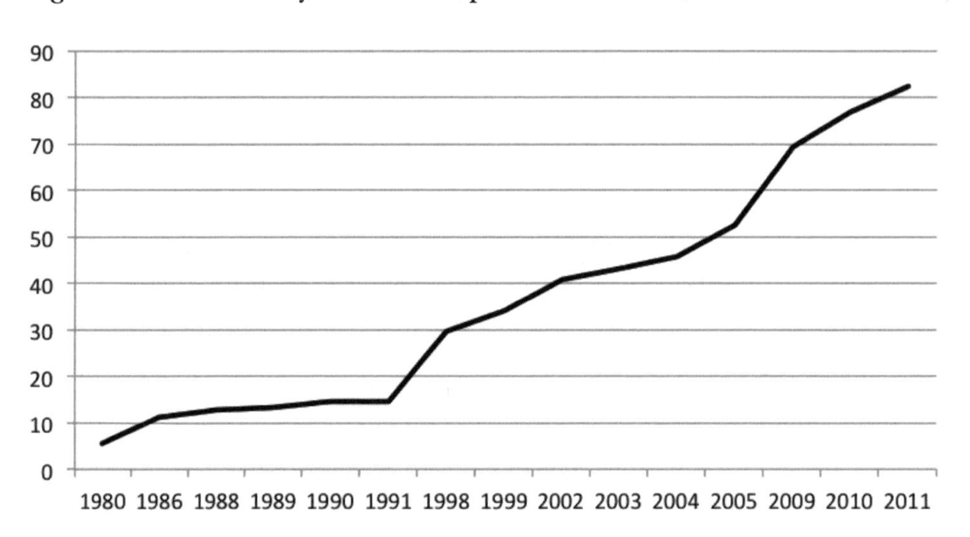

Source: U.S. Energy Information Administration (2014)

A number of environmental initiatives have been adopted to address certain issues. For example, plastic bags are set to be phased out gradually (Litter from plastic bags is responsible for nearly half of all nomadic camel deaths. In 2013, the UAE's residents consumed 11 billion plastic bags according Ministry of Environment and Water (2013). Furthermore, according to Waste Management Department of Dubai Municipality, the Emirate of Dubai alone uses 1.9 million bags a year. Therefore, the Abu Dhabi and Dubai's municipalities have organized campaigns that raise awareness about the negative consequences associated with using plastic bags, and encouraged the use of fewer plastic bags.

Realizing the need and importance of a safe, clean and efficient energy source, the UAE has invested significantly in creating the necessary infrastructure for nuclear energy production. Two nuclear energy plants are currently underway, and the first one will be operating in 2017. The Emirates Nuclear Energy Corporation (ENEC) declared that by 2020, it is expected that four plants will be online which will approximately produce 25% of the UAE's energy needs, and save up to 12 million tons of carbon dioxide emissions each year (The National 2013). It is also expected that the nuclear energy programme will generate electricity at a cost that is one-third the cost of traditional fired power plants.

Reducing carbon emissions and transition to low- carbon economy is one of Dubai and the UAE's top strategic objectives. The Dubai Supreme Council of Energy and Dubai Carbon Centre of Excellence (DCCE) signed an MoU to work together to come up with a strategy to curb carbon emissions in Dubai. Part of Dubai's energy strategy is to create the Mohammed Bin Rashid Al-Maktoum Solar Park that will generate approximately

1,000 megawatts of power relying on solar energy. Furthermore, during the World Green Economy Summit, Dubai signed its 2014 Declaration to transform its economy to green one that reduces significantly carbon emissions by five million tons[5]. This entails forcing government and private sector entities to adopt strict carbon emissions quotas in the near future.

4. Social Development Indicators

One of the most remarkable achievements of the UAE since its establishment in 1971 is the unprecedented social development it has witnessed in less than four decades. The welfare system the UAE provides its citizens with is unmatched globally. The contributions of the UAE to the welfare and social security of its citizens are mirrored in the policies and systems they set up for that purpose. The UAE generously provides free education and healthcare systems to nationals. In an August 2008 interview with the American TV program *60 Minutes*, Sheikh Mohammed Al Maktoum said that he wanted "the best health care systems, the best education and the best life style for his people." (*60 Minutes, 2008).* The UAE's focus on facilitating access to education, especially for women, led to considerable improvements in women's access to education and other social rights. A total of 85 percent of UAE women were illiterate in 1971; by 2005 that number had fallen to just 7.6 percent. Today, female college graduates outnumber their male counterparts: 70.8 percent of students attending university are now female (World Bank 2013).

The UAE has provided a safe and prosperous environment for its citizens and also expatriates. The economic and social opportunities that the government provides have attracted people from all the corners of the globe. Population growth has been substantial in the past 40 years of the UAE's history. As indicated in the World Banks' Data Bank, population growth was approximately 7% in late 1960s before the creation of the UAE's federation system.

Compared to other countries of the GCC, UAE has attracted far more people who have come to the country to live and work. 2009 witnessed a noticeable fall in population growth from around 13% to 8% due to the consequences of the financial crisis on the economy that led to the departure of a lot of expatriates (Dubai's Statistics' Office 2013). The Dubai phenomenon is unlikely to be replicated *in toto* by other cities or regions, but as a grand experiment it nonetheless offers some lessons for other Gulf states.

The government's health and education expenditures are substantial percentages of GDP. Government spending on public health care increased from 7.7 percent of government expenditures in 2001 to 8.9 percent in 2009—almost 3 percent of the GDP. Moreover, the government spends considerably more on education, which in 2009 amounted to 23.3 percent of overall government expenditures (World Bank 2013).

5. Developmental and Sustainability Challenges

5.1. Economic Challenges

The Emirati government, as other governments all over the world, suffered under the global financial crisis that affected oil and property prices. The extent to which the UAE's and Dubai's economies were battered by the financial crisis beginning in late 2008 is well known. The real estate, finance, tourism, and trade sectors were the most affected by the crisis. Dubai had to turn to Abu Dhabi for two $10 billion bond issues in 2009. By comparison, bailouts were also required for Portugal, Iceland, Ireland, Greece ($146 billion), and Spain at around this time. After the economic shocks, the property market in Dubai fell by around 30 percent in value, and the economy was still over-leveraged in 2011.

Although difficult to quantify, Dubai's reputation among foreign investors was hurt. Much of the development was built on speculation, with retail investors buying units before construction began. Those investors are spread out throughout the region, Europe, and Diasporas around the world. Dubai's judicial and regulatory systems have struggled to keep up with the backlog of cases resulting from business disputes.

As discussed in more depth in section 5.2, since 2011, the UAE's consumption of natural gas has exceeded the local production as shown in figure 4 below. This is a result of the rapid industrialization that the country has witnessed in the past 20 years and also the use of natural gas to produce energy. This makes the UAE an importer of natural gas from Qatar from the Dolphin natural gas pipeline that provides gas to the UAE and Oman for extremely low prices that are below the prevalent international market rates. (Al Ferra and Abu-Hijleh 2012).

Figure 4 **Gas Production vs. Consumption in the UAE**

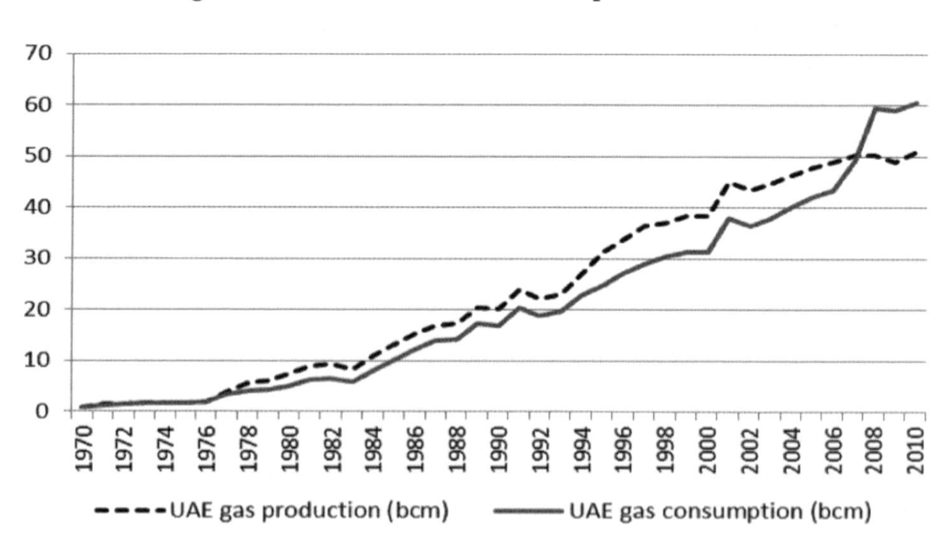

Source: U.S. Energy Information Administration (2014)

One of the domestic challenges faced by the UAE's economic development is the presence of sharp levels of disparity in each Emirate's economic performance compared to their respective shares in the country's overall GDP. This significant difference in economic development contributes to a north (underdeveloped) and south (developed) dichotomy. As shown in Table 1, Abu Dhabi and Dubai have been the major actors in the UAE economy; in 2009, their GDP distribution reached 60 percent and 30 percent, respectively. Sharjah contributes much less than Abu Dhabi and Dubai to the national GDP, but considerably more than other northern Emirates.

***Table 1* GDP Distribution among the seven Emirates**

	2001	**2002**	**2003**	**2004**	**2005**	**2006**	**2007**	**2008**	**2009**
Abu Dhabi	**48.9**	48.4	50	53.6	57.8	60.3	57.5	61	60.1
Ajman	**1.2**	1.2	1.3	1.3	1.3	1.3	1.3	1.2	1.4
Dubai	**40.4**	40.9	39.7	36.6	31.9	29.5	32.7	29.7	29.6
Fujairah	**1.1**	1.2	1.1	1	1	1	1	0.9	0.9
Umm Al-Quwain	**0.3**	0.3	0.3	0.3	0.3	0.3	0.2	0.2	0.2
Sharjah	**5.9**	5.9	5.7	5.5	6.1	5.9	5.7	5.6	6.1
Ras Al-Khaimah	**2.1**	2.1	1.9	1.8	1.7	1.6	1.5	1.4	1.6

Source: Department of Economic Statistics - National Accounts Division (2013)

Northern Emirates, such as Umm al-Quwain, contributed as little as 0.3 percent in 2001, which dropped to 0.2 percent since 2007. Similarly, the contributions of the Emirates of Fujairah and Ras al-Khaimah have not exceeded 1 percent and 2 percent, respectively. Hence, in order to achieve balanced contributions to the national GDP and decentralize economic development, it is imperative for the federal government to consider diversifying its investment plans within other Emirates. This would even out economic development among the seven Emirates and create equal opportunities among Emiratis.

5.2. Energy Issues

The backbone of the Emirati economic expansion has been the natural gas molecule, as the country focused its development on the hydrocarbon industries as the basis for its industrial production growth. Yet, at the same time, the UAE sought rapid economic diversification in order to become a modern economy. While the country is blessed with prodigious amounts of natural gas, the rise in Emirati natural gas demand strained its ability to fuel the economy through indigenously produced gas. The rise in gas demand came in the wake of increased gas consumption in the downstream value-added industries and the power-generation sector.

The UAE holds the seventh largest oil and the seventh largest gas reserves in the world, and it was also the seventh largest oil producer in 2011 with production of 3,096 million bbl/d (Eia.gov, 2011). The UAE sits on 97.8 billion barrels of proven oil reserves with a reserves-to-production (R/P) ratio of nearly 81 years, which implies that the country's current production could be sustained in the long run.

The UAE also has immense amounts of natural gas. It possesses 6.09 TCM (215 TCF) of proven natural gas reserves, accounting for 3.1 percent of the global total (Eia.gov, 2014). Yet, due to the breakneck industrialization and economic expansion, as of late, the gap between production and consumption has widened. Shortages transformed the UAE from a traditional natural gas exporter to an importer of LNG and natural gas via the Dolphin pipeline from Qatar.

The abundance of energy resources in the UAE brought enormous wealth to the country. GDP has doubled in the past decade as average growth approached 5.8 percent per annum (Imf.org, Dargin, 2014). Most of the economic development has come on the back of an increase in oil and gas output. In 2010, the energy sector comprised 32 percent of the total GDP and 76 percent of all the fiscal revenue (Darbouche, 2012). Yet, rising GDP has been coupled with expanding domestic natural gas consumption, which has not only produced widespread gas shortages, especially in the energy-poor Northern emirates, but also impacted the ability to sustain gas exports and collect the associated foreign revenue. More specifically, natural gas demand has been rising by 6.8 percent per annum for the past ten years reaching 62.9 BCM in 2012 (Bp.com, 2012). In 2008 alone, gas consumption rose by more than 20 percent, which was a stark example of the reoccurring phenomenon of consumption exceeding domestic production.

There are multiple reasons for the rapid increase in domestic natural gas demand. Firstly, along with the GDP growth, the UAE experienced a population boom in the last decade coinciding with a youth bulge that compounded already high consumption rates. Since 2001, the UAE's population more than doubled and is expected to continue to grow by 3 percent per annum for the next five years to reach 9.4 million people by 2017 (Dargin, IMF, 2014). Other major factors driving domestic gas demand are the extended utilization of Enhanced oil recovery (EOR) techniques and increased electricity consumption. 65 percent (or 56.6 Mtoe) of the country's total energy needs are supplied by natural gas, a share that will continue to rise unless current proposals for diversification of energy supply away from oil and gas and into renewable and nuclear energy come to fruition. EOR consumed 18 BCM of the total gas production in the UAE in 2009, while electricity was by far the largest consumer of natural gas outside of the resource-extracting industry with 28.3 BCM in 2009 (Dargin, 2008). Another 10-12 BCM is consumed by the industrial sector, which has been growing by an annualized average of 7 percent in the last decade.

With IMF economic growth projections of 3 percent over most of the next decade, Emirati gas demand will likely continue to grow (Dargin, IMF 2012). However, the ultimate gas growth rates will depend on the ability of the Emirati government to implement its *2030 Economic Vision* program, which aims to diversify the Emirate's economy by raising the share of non-oil output from around 50 percent to 64 percent of GDP. Abu Dhabi plans to invest approximately $160 billion until late 2014 on development projects, including constructing airports, energy terminals and new industrial cities to host more than 130,000 residents and the expansion of electricity generation capacity. Additionally, the UAE is committed to job creation for its nationals, which will require further industrial expansion. The federal government, hence, aims to boost industrial investment to an annual average of 25 percent by 2025. Moreover, the UAE focused on the expansion of its petrochemical sector by striving to expand the output of its flagship company, Borouge, to 4.5 million tons by the fourth quarter 2014 (Neuhof, 2012).

All of the above-mentioned factors will further boost natural gas demand in the next decade. In a business-as-usual scenario, gas demand will likely continue to grow by approximately 7 percent per annum until 2020 as major governmental investment in infrastructure project and industrial development begins (Kumar 2010). The foregoing means that gas demand could reach 115 BCM per year in 2020, almost doubling current annual consumption as shown in table 2.

Table 2 **Supply/Demand Projections – 2020**

Year	**Production (BCM/y)**	**Consumption (BCM/y)**
2000	38.4	31.4
2011	51.7	62.9
2015 (est.)	64.6	82.5
2020 (est.)	78.1	115.7

Source: U.S. Energy Information Administration (2014)

Meanwhile, unless there is a significant increase in production, gas supply is forecast to grow at slower rates, thereby expanding the gas deficit. Nonetheless, associated gas production is expected to increase as the UAE implements its oil production expansion plans to ramp up oil production up from 2.8 million barrels per day to 3.5 million by 2018. But, while associated gas demand will increase, EOR demands will likely increase as well in order to coax additional oil from maturing fields.

There will be an increased associated gas production from the Integrated Gas Development (IGD) and Hail Gas Development Projects that are designed to process approximately 27 BCM from the Umm Shaif and Habshan oil fields by 2014. Moreover, there is an expected 12.9 BCM increase in non-associated natural gas production as several fields come online by 2014/2015, including, as mentioned above, the giant Shah sour gas field and smaller gas developments in the northern emirates.

Due to the continued demographic growth and the demand pressures created by industrialization and economic growth, it is estimated that unless decisive action is undertaken, the UAE could experience a natural gas shortfall of 37 BCM by 2020. Yet, it must be kept in mind that these demand/production figures only take account of power demand and industrial use and exclude natural gas volumes for enhanced oil recovery and LNG exports.

But the picture is not completely bleak, for example, during a five year period between 2003-2008, the UAE was able to increase its natural gas production by nearly 5.4 BCM (from 44.8 BCM - 50.2 BCM) to meet the strong growth in local demand. The massive surge in funding during early to late 2000s allowed the UAE to rapidly increase its natural gas production, and to maintain its position as the fourth largest Arab natural gas producer after Qatar, Algeria and Saudi Arabia. What the preceding illustrates is that when far sighted proactive policies are undertaken, energy challenges can be met head-on.

Additionally, the UAE is well-suited to take advantage of its historically close relationship with IOCs, which are able to lend their expertise to development of complex non-associated natural gas fields, as well as provide LNG to the local market. Therefore, for the future, the UAE will have to remain flexible and utilize a dynamic energy mix able to bring in strong international partners for domestic gas development, as well as leverage the expanding global LNG market and implement energy efficiency programs. If these policies are undertaken judiciously, then the UAE will continue to be able to fuel its rapid economic development for the long-term.

5.3. Environmental Challenges

There are many challenges in front of achieving environmental sustainability in the UAE. The World Wildlife Fund reported in 2010 that Qatar and the UAE have the world's highest CO_2 footprint per capita. The UAE requires energy for cooling and for desalination of water supplies. These combined factors resulted in it being listed as having the highest ecological footprint in the world at 10.68 global hectares (gha) per person (WWF, 2010). The main reason behind the dramatic rise of the ecological footprint is the phenomenal economic growth experienced in the UAE that proceeded in parallel with population growth. The industrial revolution in the UAE began right after the independence of the country and the creation of the federation. Figure 5 illustrates the magnitude of Carbon Dioxide Emissions in the UAE that jumped from 150 million metric tons in 2005 to more than 220 million in 2010. This resulted in fast-growing demands for resources and

Figure 5 **Total Carbon Dioxide Emissions from the Consumption of Energy (Million Metric Tons)**

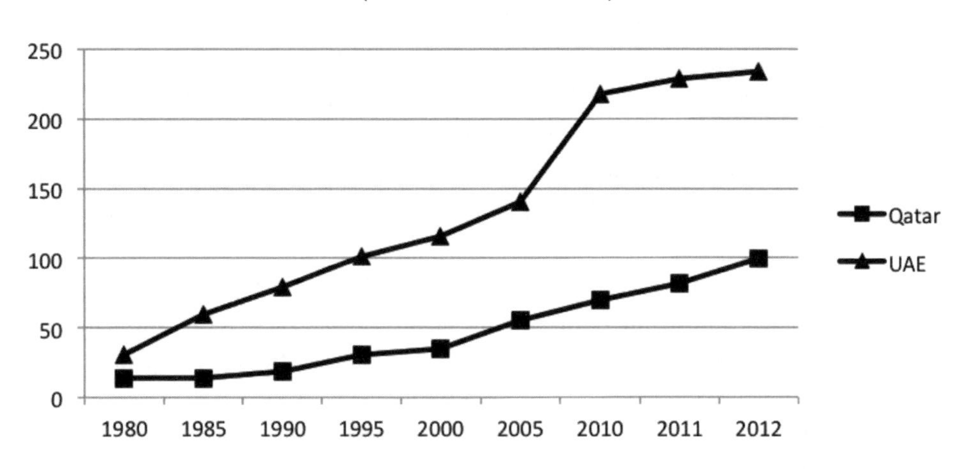

Source: U.S. Energy Information Administration (2014)

electricity, utilization for cooling, and the tremendous amount of cars fueled with gasoline subsidized by the government. These factors contributed to the rise of CO_2 emissions in an unprecedented scale. Since 1994, CO_2 emissions have increased by 64 percent (Chaar and Lamont 2010)

Dubai has both the highest per capita water consumption in the world and the most expensive water. Daily consumption rate of 350 liters a person – 100 liters more than the global average. The expensive price of water is due to its extraction costs. Around 72% of UAE's water comes from the groundwater which is extracted twenty three times its recharge rate (The National 2012)[6]. Only 25 percent of the UAE's fresh water does not require desalination or similar treatment; 25 percent is desalinated from the ocean, while roughly 50 percent of ground water is saline and requires "treatment." (Abou Elseoud et al. 2013, Shadi 2014).

There are also a variety of other environmental challenges afflicting Dubai. These range from loss of habitat for plants, animals, and sea life and the resulting decline in biodiversity, to the need to expand its capacity for treating sewage, litter, loss of limited forests, long-term climate change concerns and collapse of fisheries.

5.4. Social and Demographic Challenges

One of the main development challenges faced by the UAE is the low ratio of native Emiratis to immigrants, the latter whom comprise more than 80 percent of the overall UAE population (Dubai's Bureau of Statistics, 2013). This not only poses developmental challenges, but also causes an overreliance on foreign expertise and talent. There are three major challenges that the UAE faces in terms of its demographic growth:

First, local's population growth has been declining since 2001, dropping from an average of 5 percent in 2001 to less than 3 percent in 2009. The World Bank Data Bank categorizes population growth into 3 main age categories: 0-14, 15-64, and 64 and above. The vast majority of the current local UAE's population, according to the World Bank data, is between the ages of 15 to 64; this means that the population is capable of participating in the labor market and contributing to the overall development of the country. However, their skills do not necessarily match the requirements of the labor market. Furthermore, it is difficult for many nationals to compete with highly skilled expatriates.

Second, the UAE has a lack of sufficient and reliable demographic data on national identity development. The nation needs baseline demographic data to fully understand how the characteristics, attitudes, values, and norms of Emirati identity are being shaped and sustained by such policies.

Third, there is a critical need for evaluation of the specific mechanics of Emiratization policy implementation, as there is limited knowledge of how participation in the policy is specifically shaping the local values and norms, particularly in the public sector. The UAE has implemented numerous Emiratization policies that aim to enhance the representation of the local people in both the private and public sector. However, the effectiveness of these policies is questionable in the absence of rigorous methods to evaluate the success of these policies in addressing the low presence of locals in the labor force.

5.5. Cultural Challenges

A key challenge faced by the United Arab Emirates, and Dubai in particular, as with other GCC countries, is the risk of losing national and cultural identity. As stated above, expatriates form more than 80 percent of the UAE's population (more than 90 percent of Dubai's) and their presence could significantly test the cultural bonds and social norms of the citizens.

National identity has been described as a multidimensional concept, being controversial among various social scholars and with a variety of frameworks existing to conceptualize its definition. Indeed, the term "national identity" is defined as the depiction of an individual that is determined by gender-related, territorial, cultural, social, religious, ethnic, and linguistic characteristics (Smith 1991). Notably, Cohen (2004) argues that the construction, reproduction, and reshaping of identity is the crucial preoccupation of our era and defines national identity as a constructivist conceptualization of identities, which highlights its fluid and multiple nature. From a micro-sociological perspective, national identity is understood by considering individual understandings, as well as the relationship between individual identity construction and society as a whole (Baumeister 1986). Currently, the focus for defining identity has changed into more of a collective approach, considering how gender, sexuality, race/ethnicity, and class location influence individual constructions of identity (Appiah and Gates, 1995).

It is difficult to quantify the extent to which the UAE's cultural heritage is affected by the influx of expatriates to the country, but one cannot deny the tremendous changes in lifestyles and habits of the locals, the formation of new familial structures, and other changes. The Emiratis take pride in their world-class accomplishments achieved in a short amount of time and their improvements in material quality of life. Nevertheless, they are concerned about losing control over their traditional values, amid uncertainty about all of the changes that have occurred within a single generation. Interaction with expatriates fuels the locals' curiosity and interest in various ways, but can also lead to frictions. Wearing jeans, using mobile phones for chatting and networking among females and males, smoking and other typically Western habits that are not part of traditional culture are all practices that have been introduced or expanded in the past fifteen years.

6. Conclusion and Recommendations

This chapter has demonstrated that the government of the UAE has been making substantial progress in enhancing economic, environmental, social and cultural sustainability since the 1970s. Even with its challenges, the UAE is still a beacon of hope for young and forward-looking thinkers in the region. The negative lessons are well known, too, and the UAE's reputation took some bruising from the excesses of the past decade and the lack of sufficient institutional capacity to regulate them.

The UAE has adapted itself to the financial crisis of that hit its economy in 2009 and is making fiscal and organizational reforms to overcome its repercussions. Bureaucratic models and procedures are being adjusted, investments are being tailored to better fit demands, and future developmental projects are being approached in a more pragmatic way. The UAE is making noticeable progress with various e-government initiatives that aim to facilitate the interaction with government entities and simplify other processes.

Dubai is in a more critical position, since with only 6 percent of its GDP coming from oil its non-resource based economy has grown remarkably fast for 30 years, even taking into account the financial crisis. It has created a very comfortable and stable social life for its citizens and many of its residents. It simultaneously offers an enriched cross-cultural experience and challenges its traditional heritage, and it has proven the value of a culture in which reinvention is part of its identity.

The UAE is still heavily reliant on fossil fuels as an energy source (the UAE's first nuclear reactor is planned for 2017) and it has a high environmental footprint that will require significant institutional advances or technological breakthroughs in order to become sustainable, although it is making progress in this area. Despite the relentless efforts of the government to provide the best services to its citizens, there are still some challenges that it is facing. Based on the previous discussion, the chapter recommends:

The UAE should draw long term economic strategies that encompass and address both the internal and external pressures. UAE's GDP still relies heavily on oil sector. Since the world economy is witnessing substantial changes in response to political destabilization that is taking place in the Middle East and in Europe, the oil prices that are currently under US $80. This has a direct influence on the UAE's overall economic performance, and its ability to continue subsidizing oil prices locally. The UAE should strengthen its economic foundations by increasing the role of non-oil sector to stimulate sustainable growth. Furthermore, as part of its social contract, the UAE's public sector continues to be the major recruiter of Emiratis. Stimulating economic growth requires investing in entrepreneurial projects that will create jobs.

Dubai should regulate its real estate market and ensure its long term sustainability. Dubai's economy relies significantly on the global real estate market prices and conditions. The Dubai government suffered under the global financial crisis that affected its property prices. After the economic downturn of 2009, the property market in Dubai fell by around 30 percent in value, and the economy was still over-leveraged in 2011. An IMF report on the UAE states that "recent events call into question the sustainability of enhancing growth through large-scaled and highly leveraged property development" (IMF, 2013). The news of Dubai winning the Expo 2020 in 2014 resulted in the property prices skyrocketing again, making investors nervous about the possibility of another economic bubble.

The UAE's Northern Emirates have not been properly involved in the developmental economic trajectory that Dubai and Abu Dhabi embarked on. Northern Emirates, such as Umm al-Quwain, contributed as little as 0.3 percent in 2001, which has dropped to 0.2 percent since 2007. Establishing economic and investment hubs in the Northern Emirates will be a driver for job creation, improving the living standards of Emiratis in these Emirates, enhancing the quality of infrastructure, and increasing the contribution of these Emirates to the overall UAE's GDP. Encouragement of Small Medium Enterprises (SMEs) in Northern Emirates will be a significant enabler of economic growth. This could take many forms especially micro-financing and low rates' loans to new graduates.

The UAE's government should strongly consider implementing effective legal frameworks, and economic policies that help sustain economic growth in the long run. These legal frameworks and policies will foster trust among foreign investors, protect the rights of expats in dispute with local investors, and encourage more foreign direct investment opportunities to the country. Since the UAE wants to label itself as the investment hub of the MENA region, lenient investment policies will create more opportunities for foreigners to live and work in the country.

The government should utilize alternative energy sources to reduce its use of natural gas for electricity generation. The backbone of the Emirati economic expansion has been the natural gas molecule, as the country focused its development on the hydrocarbon industries as the basis for its industrial production growth. Yet, rising GDP has been coupled with expanding domestic natural gas consumption which has exceeded the local production. Ensuring that the four nuclear reactors are functional on time, will at least guarantee that long term efficient energy sources are secured.

The government should continue its commitment to reduce CO2 emissions. Since 1994, CO_2 emissions have increased by 64 percent. The UAE requires energy for cooling and for desalination of water supplies. These combined factors resulted in it being listed as having the highest ecological footprint in the world at 10.68 global hectares (gha) per person (WWF, 2010). Therefore, utilizing alternative energy sources that are environmentally-friendly is essential to gradually curb the high CO2 emissions. The government's current efforts are applaudable, but not sufficient to effectively and substantially reduce its carbon footprint while maintaining economic growth. More importantly, implementing a carbon emission quota that companies must abide by will be a milestone achievement in reducing carbon emissions.

UAE should invest in research based policies especially in areas related to social development and demographic growth. This will provide better understanding of the actual social challenges that UAE faces. The currently available data and research do not provide reliable recommendations or directions to policy makers on how to tackle its demographic and social challenges. The Emiratis' population growth is not growing sufficiently to provide the local workforce that is necessary for sustainable economic and social growth. Hence, reliance on the experience foreign expats who continue approximately 80% of the UAE's population will continue for the coming decades.

Since the overwhelming majority of the UAE's population is made of expats, the government should strongly consider investing its strengthening awareness about its cultural and national identities. This can be further preserved and protected by providing a platform for the Emirati youth to learn about their cultural identity to be able to pass it into the coming generations. This could perhaps be best achieved by including courses of Emirati heritage and culture in the educational systems. Furthermore, organizing cultural events at educational and professional institutions can be a reliable instrument to educate Emiratis about their culture.

Notes

1 Special thanks to Mohammed Fathi for his excellent research assistance in this chapter's data collection.
2 http://www.uaeinteract.com/docs/President_issues_decree_abolishing_Federal_Environmental_Agency/37795.htm
3 http://www.cwm.ae/pdf/Federal%20Law%20no.24%20of%201999.pdf
4 http://gulfnews.com/business/economy/uae-power-capacity-outpaces-demand-1.1068506
5 http://wges.ae/outcomes-wges_pdf_page.php
6 http://www.thenational.ae/news/uae-news/environment/water-consumption-a-dubai-family-turns-off-the-taps-for-a-weekend

References

Abou Elseoud, A. and Matthews, M. (2013). Transitional water management in the Arab region. *Environmental Development*, 7, pp.119-124.

AlFarra, H. and Abu-Hijleh, B. (2012). The potential role of nuclear energy in mitigating CO2 emissions in the United Arab Emirates. *Energy Policy*, 42, pp.272-285.

Bp.com, (2012). *Statistical Review of World Energy 2012 | About BP | BP Global.* [online] Available at: http://www.bp.com/en/global/corporate/about-bp/energy-economics/statistical-review-of-world-energy.html.

Brandon, P. (2012). Sustainable development: ignorance is fatal--what don't we know?. *Smart and Sustainable Built Environment*, 1(1), pp.14-28.

Byrch, C., Kearins, K., Milne, M. and Morgan, R. (2007). Sustainable "what"? A cognitive approach to understanding sustainable development. *Qualitative Research in Accounting & Management*, 4(1), pp.26-52.

Catenazzo, G., Epalle, A., Fragni`ere, E. and Tuberosa, J. (2010). Testing the impact of sustainable development policies in Canton Geneva. *Management of environmental quality: an international journal*, 21(6), pp.845-861.

Colvin, J., Blackmore, C., Chimbuya, S., Collins, K., Dent, M., Goss, J., Ison, R., Roggero, P. and Seddaiu, G. (2014). In search of systemic innovation for sustainable development: A design praxis emerging from a decade of social learning inquiry. *Research Policy*, 43(4), pp.760-771.

Darbouche, H. (2012). Issues in the pricing of domestic and internationally-traded gas in MENA and sub-Saharan Africa. *Oxford Institute for Energy Studies*, pp.1-37.

Dargin, J. (2008). The Dolphin Project: The Development of a Gulf Gas Initiative. *Oxford Institute for Energy Studies*, pp.1-61.

Dargin, J. (2014). *World | Data.* [online] Data.worldbank.org. Available at: http://data.worldbank.org/region/WLD.

Eia.gov, (2014). *International Energy Statistics - EIA.* [online] Available at: http://www.eia.gov/cfapps/ipdbproject/iedindex3.cfm?tid=5&pid=53&aid=1&cid=regions&syid=2011&eyid=2012&unit=TBPD.

Eia.gov, (2014). *United Arab Emirates - U.S. Energy Information Administration (EIA).* [online] Available at: http://www.eia.gov/countries/country-data.cfm?fips=TC.

El Chaar, L. and Lamont, L. (2010). Nourishing green minds in the land of oil. *Renewable energy*, 35(3), pp.570-575.

Emblemsvaag, J. (2013). How economic behavior can hamper sustainable development. *World Journal of Science, Technology and Sustainable Development*, 10(4), pp.252-259.

Government of Abu Dhabi, (2008). *Abu Dhabi Economic Vision 2030 report.* Economic Vision 2030. Abu Dhabi: Abu Dhabi Council For Economic Development.

Imf.org, (2014). *World Economic Outlook Database April 2014.* [online] Available at: http://www.imf.org/external/pubs/ft/weo/2014/01/weodata/index.aspx.

Kazim, A. (2010). Strategy for a sustainable development in the UAE through hydrogen energy. *Renewable Energy*, 35(10), pp.2257-2269.

Kumar, H. (2010). UAE gas demand rises 7% yearly. *Gulf news.* [online] Available at: http://gulfnews.com/business/oil-gas/uae-gas-demand-rises-7-yearly-1.628276.

Mondal, M., Kennedy, S. and Mezher, T. (2014). Long-term optimization of United Arab Emirates energy future: Policy implications. *Applied Energy*, 114, pp.466-474.

Moreno Pires, S., Fid'elis, T. and Ramos, T. (2014). Measuring and comparing local sustainable development through common indicators: Constraints and achievements in practice. *Cities*, 39, pp.1-9.

Neuhof, F. (2012). Petrochemical expansion to fuel 'quality jobs' for the UAE. *The National.* [online] Available at: http://www.thenational.ae/business/industry-insights/energy/petrochemical-expansion-to-fuel-quality-jobs-for-the-uae.

Patlitzianas, K., Doukas, H. and Psarras, J. (2006). Enhancing renewable energy in the Arab States of the Gulf: Constraints & efforts. *Energy Policy*, 34(18), pp.3719-3726.

Pesqueux, Y. (2009). Sustainable development: a vague and ambiguous "theory". *Society and Business Review*, 4(3), pp.231-245.

Ramirez, G. (2012). Sustainable development: paradoxes, misunderstandings and learning organizations. *Learning Organization, The*, 19(1), pp.58-76.

Reiche, D. (2010). Renewable Energy Policies in the Gulf countries: A case study of the carbon-neutral "Masdar City" in Abu Dhabi. *Energy Policy*, 38(1), pp.378-382.

Sgouridis, S., Griffiths, S., Kennedy, S., Khalid, A. and Zurita, N. (2013). A sustainable energy transition strategy for the United Arab Emirates: Evaluation of options using an Integrated Energy Model. *Energy Strategy Reviews*, 2(1), pp.8-18.

Swanson, L. and Zhang, D. (2012). Perspectives on corporate responsibility and sustainable development. *Management of Environmental Quality: An International Journal*, 23(6), pp.630-639.

UAE Interact. Available at: www.uaeinteract.com, (Accessed on 15 A*pril 2014)*.

Vidican, G., McElvaney, L., Samulewicz, D. and Al-Saleh, Y. (2012). An empirical examination of the development of a solar innovation system in the United Arab Emirates. *Energy for Sustainable Development*, 16(2), pp.179-188.

3

Environmental Performance and Public Governance in the Gulf Countries: The Emergence of Strategic-State Capabilities

Paul Joyce

1. Introduction

The Gulf governments have very publicly expressed their concerns about the environmental dimension of national development and seen environmental challenges as something that had to be addressed by national development. In contrast to the aspirations of governments, the record of the Gulf countries, with one clear exception, was very poor in terms of environmental performance over the period 2000 to 2010.

The aimof this chapter is to better understand the link between the design and capabilities of public sector institutions in the Gulf countries and their environmental development. In 1997 the International Bank for Reconstruction and Development and the World Bank produced a report, the *World Development Report 1997, in which it was claimed that there was a widespread need, in many countries, to close the gap between what was expected of the state and the capabilities of the state to act. For more than 20 years there have been attempts to rethink and reinvent governments (Osborne and Gaebler, 1992) to increase their effectiveness.*

One of the implicit issues for this chapter is to see if a country'sperformance on the environment was linked to economic development and growth. Government effectiveness might be expected to show up both as better economic growth and better performance on the environment. This issue will be addressed by looking at economic growth in recent years and seeing if its determinants were different to, or the same as, those for environmental performance. The case of the United Arab Emirates will be especially interesting on this point, since it was the main exception to a poor record on environmental performance.

The intended outcomes of this chapter are the clarification of how public governance has developed in the Gulf States, before and after the international financial crisis of 2007-2009, and to see if the poor environmental performance can be understood in relation to public governance. But, first, the chapter considers the evidence on environmental performance.

2. The Environmental Performance of the Gulf States

As was noted in the introduction, the governments of the Gulf States were concerned about the natural environment. But how well were they doing? One authoritative source of comparative data on environmental performance is the Environmental Performance Index (EPI) report (see Emerson et al. 2012). The EPI and the Pilot Trend Environmental Performance Index (Trend EPI) use a range of performance indicators to assess the environmental public health and ecosystem vitality of countries. In the 2012 report, the Trend EPI focuses on changes in performance between 2000 and 2010. The report's authors, speaking about all 132 countries included in the index, suggested that many countries had improved their environmental performance over the period, although they expressed concern about the continuing global rise of greenhouse gas emissions and the challenge countries faced in achieving a sustainable emissions trajectory.

The data presented in Table 1 is taken from the 2012 report and shows the results for five of the six Gulf States. The data suggests that the United Arab Emirates had managed to achieve a big improvement in its EPI performance over the ten years (appearing as ranked 27 in the Trend EPI data), and Oman's result was encouraging (with a Trend EPI rank of 80 as against an EPI rank of 110) but the data showed that Saudi Arabia and Kuwait were not ranked highly in terms of the trend EPI. In fact, in relation to all 132 countries, they were among a small group of countries with the most negative trends in environmental performance (ranked 130 and 131 respectively in the Trend EPI Rank) and only Russia had a worse trend.

Table 1 **Environmental Performance of the Gulf States**

Country	Environmental Performance Index (EPI) Rank	Trend EPI Rank (2000-2010)
Kuwait	126	131
Oman	110	80
Qatar	100	121
Saudi Arabia	82	130
United Arab Emirates	77	27

Source: Data from Emerson et al. (2012)

The EPI Report can be triangulated against World Bank data to deepen our understanding of the environmental performance of the Gulf countries. Table 2 displays two indicators: CO2 emissions and renewable internal freshwater reserves. In terms of CO2 emissions (metric tons per capita) there were three countries which had CO2 emissions that were declining over the period 2000 to 2010: Bahrain, Qatar and UAE. There were also three countries where CO2 emissions were rising on a per capita basis: Kuwait, Oman, and Saudi Arabia. Turning to the trends in renewable internal freshwater reserves per capita, all five of the countries for which there was published World Bank data experienced a decline in the availability of renewable internal freshwater reserves. Over the decade covered (2002-2011) the reductions in these water reserves were substantial in percentage terms – ranging from 21 percent in the case of Saudi Arabia to 67 per cent for Qatar. Of course, the governments of the Gulf were aware of difficulties in terms of water availability.

Table 2 World Bank Data on CO2 Emissions and Renewable Internal Freshwater Reserves

	Bahrain	Kuwait	Oman	Qatar	Saudi Arabia	UAE
Year	CO2 emissions (metric tons per capita)					
2000	27.9	28.9	10.0	58.5	14.7	37.2
2004	21.3	28.9	11.4	61.6	16.6	31.0
2008	21.8	29.6	15.8	50.0	15.9	23.4
2010	19.3	31.3	20.4	40.3	17.0	19.9
Year	Renewable internal freshwater reserves per capita (cubic metres)					
2002	5.5	n.a.	606.5	88.9	110.0	46.5
2007	3.9	n.a.	544.8	48.6	92.6	25.9
2011	3.1	n.a.	462.8	29.3	86.4	16.8

Data source: www.databank.worldbank.org

So, the environmental data from the World Bank tended to back up the interpretation above of the EPI data: trends in the environmentover the decade from 2000 provided reasons for concern, including the availability of renewable freshwater reserves and also on CO2 emissions, despite the positive trends on CO2 emissions for the United Arab Emirates, Qatar and Bahrain. But why had the negative trends occurred? Did the Gulf governments have the needed capabilities to successfully address the environmental challenges?

It is worth pointing out here that the Gulf governments, whatever the extent of their capabilities, would have been potentially pulled in a number of different directions, and often at the same time.Strategic issues often place governments under contradictory pressures (Nutt and Backoff 1992). The governments would not only have had pressures on them to formulate and implement successful environmental policies, they will also have felt pressures to improve the standard of living of citizens, enhance the quality of life experienced by them, and ensure that there were plentiful job opportunities (especially for the relatively large populations of young people). They will also have had demands placed on them to uphold traditional values, and at the same time they will have had to manage the consequences of the booms and subsequent downswings in oil prices. In the face of all these issues and external pressures, the question becomes what type of public governance would enable them to be handled most effectively? And, then, what capabilities are most important for this type of governance? The next section looks at the concept of the rentier state, which has often been seen as applicable to the Gulf States. This will then be followed by a review of signs that the Gulf governments might have been reforming themselves to develop strategic-state capabilities.

3. The Rentier State

The model of the rentier state could be seen as just a technical or descriptive concept with no evaluation implied. In practice, discussions of the rentier model may seem at times to be tinged with a value judgement, because possibly of some of its perceived negative consequences, but also, maybe, because some commentators see it as based on "unearned" income. In summary, in an oil rich country the rentier model denotes a form of government in which a very large part of the income of society is controlled by the government and, speaking economically, a large part of this is seen as resulting from a big difference between the price of oil on world markets and its actual costs of production. In fact, the World Bank statisticians define "oil rent" in precisely this way. On this basis, the income of the government is seen as a "windfall" or a lucky chance, the government just happens to be in a country with an abundance of oil. The country did not have to work for it.

The contrast can then be made with countries which do not have such resources, and where the state is therefore strongly interested in policies promoting economic growth, productivity, innovation, and exporting goods and services. The implication is that in this case, the state has a strong interest in "production" - we might say it is a production state. By contrast, the problem for the rentier state is the allocation of the rent from, say, oil. In fact, some suggest that an alternative name for the rentier state is "allocation state" (Hvidt 2011).

What else is associated with a rentier state? Baghat (1998) brought out the implications for public services.He argued that oil revenues provided the Gulf States with immense wealth, which funded a generous welfare state, and citizens benefitted from subsidised or completely free goods and services. Moreover, he stated that it was because of oil revenue that the state did not need to rely on tax to pay for government services. Another consequence claimed for the rentier state is the make-up of employment opportunities for citizens; Hertog (2010) suggests that very many citizens in a rentier state are employed in the public sector (Hertog 2010).

In some accounts of the rentier state there are suggestions that the rentier state creates problems. These may be economic, cultural, and political.Hvidt (2011 p.88) mentions some economic consequences:

> "A rentier economy is an economy where the creation of wealth is centred on a small fraction of the society... When the state is liberated from the national economy in this way, it is not under pressure to develop an efficient economic basis for the country, that can rather rest on distributing (or allocating) the revenues it accrues from rents."

Hvidt (2011) also identifies cultural effects. A rentier state may foster a "rentier mentality" which is one where there is a disconnection of the "normal" link between work effort and payoff because the individual is lucky enough – by accident – to be in a country where there just happens to be a lot of oil. So, extending Hvidt's argument, it might be said that the values of society come to reflect a low valuation of work and risk taking, and then each new generation is socialised into these values so that there is a problem with the motivational patterns needed for a modern economy based on the importance of economic growth, innovation and productivity.

Another suggestion is that oil rich Gulf States may generate a special type of role in society not found so commonly in developing countries. Hertog (2010) theorises that rentier states create large numbers of people who he described as involved in brokerage, by which he meant individuals who enable others to gain access to the state's resources. He identifies negative consequences of large scale brokerage activities, suggesting that brokerage is implicated in "creating various rentier effects, including slow growth, local productivity, weak policy implementation, and weak political participation" (Hertog 2010 314)

There also seems to be arguments that rentier states have a specific effect on how the business community relate to the political system.Moore (2002), who does not agree with this assumption, summarises these arguments as follows:

> "Rentier state theory derived from study of the Gulf cases charts the creation of an "oil social compact", business abandons political involvement for state mediated profits ...and although links with the state are robust, domestic business is hardly autonomous" (Moore 2002 p. 36-7).

He considers this model of the impact of the rentier state on government-business relations to be flawed. Based on actual comparisons of government-business relations, Moore highlighted variations in practice between how different Gulf governments and businesses had actually related, with some examples of quite autonomous business elites that worked in partnership with government on policy making.

It is not difficult to see how models of the rentier state can end up assuming that the rentier state operates in an autocratic or neo-patrimonial manner. In boom times, when the price of oil is high, and oil rent revenues are high, the government is providing a high standard of living, little or no taxes, and it might be assumed that the consent of citizens to a paternalistic and traditional pattern of governance is quite unproblematic. In the 1990s, after a long period of oil prices being relatively low, commentators might speculate that this could create problems for political stability. But with oil prices booming, then the integration of citizens might be perceived as secured through state funded or state subsidised consumption.

One response to the assumption that oil rich rentier states are paternalistic and autocratic may be to say that there are forms of interaction between rulers and the citizens, very different from a Western style of parliamentary democracy, but nevertheless in keeping with traditional values in the Gulf States. And that these interactions ensure continuity between the needs and aspirations of citizens and the decisions of their rulers.A different response is to question the viability and suitability of a Western model of democracy as the right form of public governance for modernisation in the 21st century. Indeed the World Development Report 1997, referred to earlier, gave as one of the reasons for concerns about the effectiveness of current models of the state was the important role of the state in what it termed "miracle" economies. Khonker (2011) seems to be making a similar point when he emphasises the possible lessons for the Gulf of the success of Singapore. As he (2011 p. 306) puts it: "Similarly, the conclusion that a liberal democracy is the only trajectory of modernization is also simplistic.... Singapore's remarkable economic growth and its success in creating a safe, stable society receive much attention in the Gulf.... a tacit acceptance of the position that there are multiple roads to modernization seem to have gained grounds. A careful scan of the world will reveal that the path to modernity is not single but multiple". Other examples worth attention might be Malaysia and Turkey, and it could be argued that these might be more relevant to the cases of Saudi Arabia and Oman than Singapore, which could be seen as more relevant to Dubai and Bahrain.

4. The Rise of the Strategic and Enabling State

Starting with the work of Osborne and Gaebler (1992), through the *World Development Report 1997, and subsequently there was a growing conviction that modern states should be offering strategic and enabling public governance. This occurred through best practice sharing at international conferences and through the work of policy experts in individual countries (e.g. in the UK over the period 2000-2007). The financial crisis of 2007-9 seemed to encourage an even stronger appreciation of the urgency of this public sector reform agenda. In November 2012 the OECD held a Global Forum on Public Governance, titled "Better Governance for Inclusive Growth", attended by policy makers, academics and others from 63 countries. The OECD's summary of the Forum suggested:*

> "The financial crisis revealed serious weaknesses in the governance and regulatory structures necessary for promoting a level playing field and inclusive input to policies. Good governance and quality public services are crucial for competitiveness and growth in both developed and developing countries. This issue is central to development strategies..."

The OECD in the years immediately after the financial crisis carried out a number of public governance reviews, at the heart of which was a model of increasing the capabilities of government by developing strategic-state capabilities.

We can infer from this outline history that one of the key propositions in contemporary public sector reform is that with the right type of public governance and with the necessary state capabilities governments can deliver sustainable economic growth. Moreover, it is evident from this literature that the type of public governance being promoted is one in which government works for sustainable growth on the basis that environmental goals for sustainability, are integral to the definition of sustainable growth. It is not just a matter of economic sustainability. It is public governance conducted in a strategic and focused way, engaging with the business community and with citizens, being responsive to public needs, and providing high-quality public services.

The first piece of evidence that the Gulf States have in recent years been attempting to develop strategic-state capabilities was their formulation of ambitious aspirations for modernisation and progress in long-term vision statements and associated strategic plans and development plans. The Omani government produced the first of the long-term vision statements.It held a conference in Muscat in 1995 to create a Vision for 2020. The second one was produced by Saudi Arabia. This was the result of a symposium on the future vision for the country in Riyadh in 2002. Its long-term vision statement was called Vision 2025. Two more Gulf States produce vision documents in 2008; Qatar created its National Vision for 2030 and Bahrain launched an Economic Vision for 2030 in late 2008. The last two were UAE with UAE Vision 2021 and Kuwait with Vision for 2035. UAE Vision 2021 was launched in February 2010 at the closing of a cabinet meeting.In the same year, 2010, the Kuwait Supreme Council for Planning and Development (SCPD) issued the Kuwait vision statement and strategic objectives (2035), and mid-term development plan (2010-2014).

It is also evident from web sites and documents that the long-term vision statements were in each case the target for a strategic management system. Oman intended to use its Vision to direct its system of five-year development plans, which were supposed to be formulated so that they were a means of delivering the Vision.

For example, the seventh of Oman's five-year development plans, covering 2006-2010, set a national target for economic growth, aimed at enhancing diversification of the economy, aimed at encouraging local and foreign private sectors, and so on. Of course, this system also meant that there could be an emergent aspect to the national priorities.

Saudi Arabia's Vision 2025 was to be supported by five-year Development Plans. Each development plan was intended to be part of a process of delivering the aspirations outlined in Vision 2025. Apparently, it was expected that the plans would use projects in different sectors to develop the economy. The Qatar Vision statement made it clear that strategic management was to be used for the realisation of the vision. A National Strategy was to be developed, with its preparation being facilitated by the General Secretariat for Development Planning, and adopting a partnership approach by consulting civil society, the private sector and government ministries and agencies. The National Strategy would provide the framework for management of the development of Qatar, with the spelling out of goals, time-bound targets, performance standards and responsibilities.

Following the launch of Bahrain's vision statement in late 2008, Bahrain's Economic Development Board facilitated the work of creating the first National Economic Strategy, which was intended to deliver the Vision.

Kuwait's case was interesting for reasons of the public nature of the difficulties encountered during implementation. It was intended that medium-term development plans would be the means of delivering the Vision. The first was designed to cover the period from 2010/11 until 2013/4 and included performance targets on key priorities, such as diversification (share of fiscal revenue from non-oil sectors) and growth (real GDP growth). The Emir of Kuwait instigated the first development plan by ratifying a law in February 2010.

This national development was seen officially as Kuwaiti government making a commitment to work through the use of strategic planning.

> "Through this plan, the Kuwaiti government aims at establishing a new ideology for strategic planning and visions to transform into a true reality, and will act as the first move towards attaining the desired vision for 2035..." (Al-Diwan Al Ameri, 2014)

Very interestingly, there was a view among some observers that part of the difficulties that were encountered in implementing the first Kuwaiti development plan was partly a result of problems in the relationship of the parliament with the planning system. This point is revisited later in this chapter as part of a discussion of strategic-state capabilities.

The United Arab Emirates (UAE) was one of the last of the Gulf countries to launch a vision statement. Just as with the other countries, the Vision statement was to be delivered through plans, in this case the UAE Government's Second Strategy, announced in 2010, and planned to be implemented from 2011 through to 2013.

To conclude this section, and summarise the policy context of the public governance developments being evaluated, it is noted tthat the following national development priorities were established through the long-term visions and plans that these six countries formulated between the mid-1990s and 2010. Firstly, there was an overarching interest in economic growth, which in some cases was to be judged by measuring increases in real GDP per capita. There was, secondly, a priority of making sure that the citizens of the country were benefitting from economic success, which might be judged in terms of standard of living, quality of life, educational and job opportunities, and so on. Thirdly, there seemed to be quite a consensus that reducing dependence on oil revenue was a top priority and that this meant a pursuit of policy of diversification. It might be usefully noted here that the growing demand for consumption by people inside Saudi Arabia meant that it could not be safely assumed that oil would always provide export income for the country; Lahn and Stevens (2011) for example, estimated that Saudi Arabia's domestic consumption of oil was rising at such a rate that all its production of oil would be needed to satisfy this home demand by 2038.Fourthly, there was a concern that the private sector should be driving the transition to a new more diversified economy and that this would require private as well as public sector investment. Fifthly, ensuring the country offered a good environment for business was a top priority for government action. It is also evident that there was an emergent awareness that Gulf countries needed to pay more attention to the environment, including water resources. Finally, it can be underlined here that leaders of the Gulf States needed a massive transformation of the functions and capabilities of their governments in order that there might be successful efforts to deliver long-term strategic visions in which the private sector was expected to be the driving force economic prosperity and growth.

The following quote from Bahrain's Economic Vision 2030 illustrates some of the summary points made, especially the focus on economic development, diversification, the importance of the private sector in driving economic development, the intended benefits for citizens, and a steering role for government:

> "We aspire to shift from an economy built on oil wealth to a productive, globally competitive economy, shaped by the government and driven by a pioneering private sector – an economy that raises a broad middle class of Bahrainis who enjoy good living standards through increased productivity and high-wage jobs. Our society and government will embrace the principles of sustainability, competitiveness and fairness to ensure that every Bahraini has the means to live a secure life and reach their full potential." (Economic Vision 2030, page 3.)

5. The Withering of the Rentier State?

Hvidt (2011 p. 88) has claimed that there was a purposeful attempt by Gulf governments to move beyond the rentier state:

> "All the GCC states have launched development plans...These development plans thus send a clear signal of the perceived need to displace the rentier state model which has dominated the economies of these states since the advent of oil and gas...".

It was evident from their vision statements and strategic and development plans (discussed earlier in this chapter) that all the Gulf states had been rethinking the role of government in economic development and in many respects appeared to be embracing the idea of government being strategic and focused and leveraging their capabilities by working with the private sector, who they envisaged would be driving economic transformation.

Recent claims by one commentator appeared to suggest that the capabilities of public governance in the Gulf had already improved, as demonstrated by how well the governments had handled the oil boom that got underway after the turn of the century. Hertog (2007) reported that the oil boom has been better managed by the Gulf governments and he suggested there were signs that public services would be improved and the infrastructure would be systematically improved.

So, where were the Gulf governments of the period 2010-2011? Were they rentier states, as some have claimed? Or had they moved on? Had they all moved on, or just some of them? What had they become? And could this be related to the variations in environmental performance we have already examined? For example, had the United Arab Emirates and Oman (to a lesser extent) developed more strategic-state capabilities given their progress in environmental performance?

The World Development Indicators from the World Bank can be used to assess the move away from a rentier state model. The first indicator is oil rents (defined as the difference between world prices of crude oil and the cost of production). The second indicator is general government final consumption, which is made up of current expenditures for goods and services and includes employee compensation. The third indicator is the tax revenue collected by the government. The fourth indicator is foreign direct investment (FDI) defined as the net inflows of investment to acquire a lasting management interest (10 percent or more of voting stock) in a business. All four of these indicators published by the World Bank are expressed as a percentage of GDP. Logically, if the rentier state is being replaced then it might be expected that oil rents would reduce as a percentage of GDP, general government consumption would decline, tax revenue would increase, and FDI would increase. The data is shown in Table 3.

Table 3 Indicators of the Move Away from a Rentier State Model

	Oil Rents (% of GDP)	General government final consumption (% of GDP)	Tax Revenue (% of GDP)	FDI (% of GDP)
Bahrain				
2000	20.6	17.6	4.2	4.6
2004	19.9	16.7	4.9	7.7
2008	26.5	13.4	n.a.	8.2
2011	19.2	13.7	n.a.	2.7
Kuwait				
2000	49.4	21.5	n.a.	0.0
2004	48.8	19.9	1.3	0.0
2008	60.7	13.4	0.9	0.0
2011	49.9	14.8	0.8	0.5
Oman				
2000	42.3	20.7	1.6	0.4
2004	37.9	22.5	2.0	0.5
2008	40.4	14.2	2.4	4.9
2011	40.2	17.2	2.2	1.1
Qatar				
2000	30.5	19.7	n.a.	1.4
2004	27.7	22.7	25.6	3.8
2008	22.7	17.9	16.0	3.3
2011	14.4	12.3	n.a.	-0.1
Saudi Arabia				
2000	40.3	26.0	n.a.	-1.0
2004	45.0	22.9	n.a.	-0.1
2008	64.3	17.7	n.a.	7.6
2011	55.5	19.4	n.a.	2.4
UAE				
2000	18.2	n.a.	n.a.	-0.5
2004	18.3	7.8	n.a.	6.8
2008	25.1	5.8	n.a.	4.4
2011	21.9	7.3	n.a.	2.2

Data source: www.databank.worldbank.org

An inspection of Table 3 suggests that generally the Gulf States ended up in 2011 more or less as dependent on oil as they were in 2000. The two exceptions to this were Saudi Arabia, which showed bigger oil rents in 2011 than in 2000, and Qatar which showed a marked downward trend in oil rents throughout the period. The one indicator that was consistent with a move away from a rentier state model (for all but one of the Gulf States) was general government final consumption, which tended to reduce as a percentage of GDP. The evidence on tax revenue was far from complete, and it can only be commented that in the cases where there was some data, that is, in the case of Kuwait and Oman, there did not appear to be a shift towards higher levels of tax revenue. In addition, the gaps in the World Bank's data on tax revenue also prevented a systematic comparison of this form of government income and the oil rent, but it is evident that oil rent tended to much greater than the relatively small amounts of tax revenue. Finally, the data on FDI failed to indicate any consistent and strong trend towards more FDI. It seemed that net flows of investment had risen, peaked and then fallen back. The peak years for FDI as a net inflow varied from country to country. The UAE's best year had been as early as 2004 when FDI had been 6.8 percent of GDP. Bahrain's FDI reached 18.4 percent in 2006 before falling away. For Oman it was 8 percent in 2007 and for Kuwait, Qatar and Saudi Arabia the peak year was 2009. In each case, after the peak year there was a tendency for FDI to fall back to low levels.

On this basis, we might conclude that the main way in which a rentier model of the state was under pressure was in terms of general government final consumption, which was clearly becoming less important in the economies of the Gulf States. The only other major shift that pointed to a move away from a rentier state was the reduction in oil rents for Qatar. On this basis, we might suggest that Qatar was the best candidate for movement towards a post-rentier state, although this is a judgement based on benchmarking the six countries with each other and does not indicate that in absolute terms Qatar was now a state no longer dependent on oil. Even in 2011, when its oil rents had diminished a lot as a proportion of GDP compared to a decade earlier, its oil rents were calculated to be in excess of 14 per cent of GDP.

6. The Rise of a Modernized State in the Gulf?

For the purposes of this analysis the chapter will concentrate on three dimensions of a modernised state based on policy thinking at the turn of the 21st century. For our purposes, a modernised state will be defined as one that has efficient and effective public services, that is responsive to citizens, and that facilitates a competitive private sector.

The concept of efficient and effective public services will be assessed using the World Bank's Worldwide Governance Indicator "Government Effectiveness". This indicator is constructed froma set of perceptions: the quality of public services, the quality of the civil service, the independence of the civil service from political pressures, the quality of policy formulation and implementation, and, finally, the credibility of the policies in terms of government commitment to them.

It is quite possible that responsiveness to the public can be manifested in a variety of ways and may indeed depend in part on the willingness and ability of political leaders to listen to citizens. It is also likely that we should not read off responsiveness to the public from the formal design of political institutions (and indeed constitutions). In this analysis we make use of Worldwide Governance Indicators published by the World Bank that include an indicator labelled as "Voice and Accountability". This indicator is also based on perceptions and is defined as perceptions of the extent to which citizens are able to take part in selecting their government and also on the presence of some liberal democratic freedoms (freedom of expression, freedom of association and free media).

The facilitation of a competitive private sector is operationalized here using the World Economic Forums Global Competitiveness Index, which is formed from twelve different elements that are categorised as basic requirements, efficiency enhancers and innovation and sophistication factors. The data, however, only covers the last part of the period being considered.

Table 4 presents the data for these three indicators and they seem to suggest a rather mixed picture, with the facilitation of a competitive business environment occurring for all 6 countries, but with a decline in voice and accountability for all of them. In respect of government effectiveness, three of them (Bahrain, Oman and Saudi Arabia) were scored as having lower government effectiveness in 2011 than they did in 2000, while the other three were all reported to have developed more government effectiveness over the same period. Qatar had also done relatively well in improving the conditions for business competitiveness keeping it in the lead of the other Gulf States in respect of the Global Competitiveness rankings.

7. Environmental and Economic Growth Performance and the Capacity of the Gulf States 2000-2011

This section of the chapter examines the evidence that there is a link between government capacity (as measured by the Government Effectiveness Indicator data from the World Bank) and environmental performance and economic performance.

Table 5 suggests that there is a prima facie case for the argument that there was a link between Government Effectiveness and trends in environmental performance in the case of the five Gulf States. Furthermore, there is a primafacie case, based on the correlation of rank orders, that it would have been possible to predict the rank order of environmental trend

Table 4 Modernised State

	Government Effectiveness (Percentile rank)	Voice and Accountability (Percentile rank)	Global CompetitivenessIndex (Ranking)
Bahrain			
2000	72	18	n.a.
2004	73	28	n.a.
2008	67	23	37
2011	69	13	35
Kuwait			
2000	52	44	n.a.
2004	61	40	n.a.
2008	56	32	35
2011	55	31	34
Oman			
2000	69	30	n.a.
2004	71	28	n.a.
2008	67	19	38
2011	63	18	32
Qatar			
2000	69	38	n.a.
2004	71	37	n.a.
2008	72	22	26
2011	75	21	14
Saudi Arabia			
2000	46	8	n.a.
2004	45	10	n.a.
2008	52	5	27
2011	44	3	17
UAE			
2000	78	32	n.a.
2004	77	26	n.a.
2008	78	22	31
2011	82	24	27

Data source: www.databank.worldbank.org; for Global Competitiveness Indexsee Hvidt (2011)

Table 5 Does Government Effectiveness Make a Difference?

	Rank Order of Government Effectiveness Average Percentile Rank 2000-2010	Rank Order of Trend EPI Rank 2000-2010	Rank order of % change in GDP per capita (constant 2005 US $) 2000-2010
Bahrain	3	n.a.	5
Kuwait	5	5	4
Oman	4	2	2
Qatar	2	3	3
Saudi Arabia	6	4	1
UAE	1	1	6

performance of the five countries based on a knowledge of their average percentile rank on the Government Effectiveness Indicator. Both Kuwait and Saudi Arabia were identified by the authors of the EPI report as among a group of countries with the most negative trends in environmental performance. These are also the two Gulf countries with the lowest average percentile ranks on the Government Effectiveness Indicators. Likewise, the United Arab Emirates had by far the most positive trend of improvement on environmental performance among the five Gulf States and it also scored the highest average percentile rank on Government Effectiveness. The only discrepancy in this argument is that Qatar, which had the second highest average percentile rank on the Government Effectiveness Indicator, did not manage to have the second most positive trend on environmental performance - it was in third position, behind Oman.

What is also apparent from Table 5 is that it appears more problematic to suggest a direct link between Government Effectiveness and the percentage change in GDP per capita over the period 2000 to 2010.(It must be acknowledged that using a per capita measure may be oversimplifying the economic growth issue and it may be important to also consider the differential experience of economic growth created by variations in income inequality between the 6 countries; at the present time no comment can be made on the existence, nature and size of any income inequality effects.) Arguably, the possibility of a simple link between Government Effectiveness and change in GDP per capita is disrupted by: Saudi Arabia with the lowest score on Government Effectiveness having achieved a very substantial percentage increase in GDP per capita, by Oman perhaps again exceeding expectations based on its Government Effectiveness score, and by the United Arab Emirates, which had the best score on Government Effectiveness doing by far the worst on change in GDP per capita (especially after 2004). If these three countries were excluded, the remaining three would still not demonstrate a simple link between Government Effectiveness and change in GDP per capita. Obviously, we might suggest that this could be caused by Kuwait doing better than might have been expected based on Government Effectiveness or Bahrain doing slightly worse than expected based on Government Effectiveness. Table 6 shows the percentage changes in GDP per capita over the period 2000 to 2010 and confirms the big disparities in the change in GDP per capita over this period.

Table 6 **GDP per capita (constant 2005 US $)**

Year	Bahrain	Kuwait	Oman	Qatar	Saudi Arabia	UAE
2000	14,992	28,571	11,855	48,667	12,838	45,969
2004	15,218	33,260	12,061	55,526	12,846	47,081
2008	14,820	34,463	15,145	52,163	15,115	31,130
2011	14,052	29,338	13,903	54,792	17,050	23,796
% Change from 2000 to 2011	- 6.3%	+2.7%	+17.3%	+12.6%	+32.8%	-48.2%

Data source: www.databank.worldbank.org

So, in real terms, the economic growth performance of Saudi Arabia, Oman, and Qatar was good over the period considered, and the growth performance of Saudi Arabia after 2004 (the year in which the long term vision was launched) was especially impressive. In contrast, both Bahrain and the United Arab Emirates had negative trends between 2000 and 2011. In the case of the United Arab Emirates, GDP per capita in real terms almost halved, with the deterioration setting in from 2004 onwards.

8. Strategic-State Capabilities

We seem to have arrived at a very puzzling conclusion about the benefits of Government Effectiveness; there is a case that Government Effectiveness was important as a cause of a good environmental performance by Gulf States from 2000 to 2010, but that there is not a clear link between Government Effectiveness and economic growth (GDP per capita increases).

A possible explanation could be that the Government Effectiveness indicator was useful for the period 2000 to 2010 when trying to explain government progress on its environmental performance because it was measuring a relevant set of state capabilities.

But a different set of state capabilities might have mattered for improving economic performance (as measured by changes in real GDP per capita) and unfortunately these different capabilities were not being measured by the Government Effectiveness Indicator. So, what might this different set of state capabilities be?

In the interests of developing a hypothesis for future research the last section of this chapter is going to be a little speculative. The starting point for the speculation is to ask what did Oman and Saudi Arabia have in common apart from being the best two performers on changes in GDP per capita over the period? Both covered big geographical areas and had relatively big oil rents (as a percentage of GDP). And what did the United Arab Emirates and Bahrain have in common apart from being the worst two performers on change in GDP per capita? They did depend on income from abroad or importing capitaland had relatively small oil rents. The oil rents point may be a spurious or misleading consideration, however, since Kuwait had relatively more oil rents than did Oman, and Qatar had even smaller, relatively speaking, oil rents than either Bahrain or the United Arab Emirates.

It is interesting to note that Oman and Saudi Arabia were early adopters of a strategic long-term approach as shown by their vision statements being the first to be launched (1995/6 and 2004). Likewise, the United Arab Emirates could be classified as a laggard in this respect because it did not launch its long-term vision statement until February 2010. So the hypothesis might be that changes in GDP per capita depended in part on strategic-state capabilities and that Saudi Arabia and Oman not only were quick to adopt strategic management but were also successful in developing the required capabilities. Conversely, the United Arab Emirates might have not only been later to commit to strategic management but might not have developed the required capabilities to be effective in using strategic management.

Some words of qualification are necessary: the speculation here is based on using the date of the launch of a long-term vision statement as roughly indicating the start of a learning process by government to develop strategic-state capabilities. This assumption may be prone to some error. For example, Saudi Arabia did have a system of development planning before 2000 although one description of the National Development Plans was that they were like shopping lists; and it has been reported that the switch to long-term thinking and planning was linked to the seventh National Development Plan which was implemented in the period 2000 to 2004. So, the development of strategic-state capabilities may have been taking place in the period just before 2000 and during the implementation of the development plan from 2000-2004.Others may find their development of such capabilities delayed or blocked even after a long term vision statement has been launched. And, in addition, we can expect strategic-state capabilities to be learnt to some degree by doing strategic management and the capacity for learning in this way may vary from one government to another, so that a government that had made an early start may not have developed strategic-state capabilities as much as another government which started later.

This word of qualification may be applicable to the cases of Bahrain and Kuwait. Their cases may just show that other factors were at work and make a simple focus on the official date of a long-term vision being launched insufficient as a proxy for the start of the development of strategic-state capabilities. So, it may be that Kuwait's government had been developing strategic-state capabilities in advance of launching its vision statement. Or, alternatively, for Bahrain's government the process of beginning to develop strategic-state capabilities lagged the launch date of the vision statement.

What strategic-state capabilities might the Gulf States have been developing? The following are some suggestions.

1. Where there are parliaments with elected representatives there is a need to develop an effective working relationship between the parliament and the government in relation to strategic plan formulation and monitoring of implementation. In fact, there were reports that progress on some aspects of implementing the 2010 vision statement of Kuwait had been stalled up until 2013 because government and parliament had been arguing over the medium-term development plans which were intended to implement the vision (QNB 2013). Effective partnership between the two can be seen as a key strategic-state capability.
2. Integrating operational plans into the delivery arrangements for long-term strategic plans. When the Saudi Arabia government was developing its Long Term Strategy in 2004 one of the things it had to do was address a pre-existing system of short-term operational plans when what was needed was a system of strategic planning linked to the long-term vision statement. This, if successful meant, developing the capabilities of civil servants for planning ahead and thinking further ahead.
3. A third and challenging capability is getting strategic plans to integrate budgeting and performance measurement into processes for delivering the long-term visions.
4. A fourth capability would be based on learning to develop partnership working with the business sector and civil society, in strategy formulation and in other matters. The idea of consultation and partnership was mentioned, for example, in the Qatar National Vision 2030 statement (page 18): "As a next step, the General Secretariat for Development Planning (GSDP), with the guidance of Qatar's Higher Authorities, will coordinate the formulation of this National Strategy, in consultation and full partnership with all stakeholders, especially civil society, the private sector, ministries and government agencies."

One implication of this line of speculation about different sets of capabilities is that the World Bank's Government Effectiveness Indicator has been measuring good government at a bureaucratic stage of a state's development. Its formal definition suggests that this might be the case. For example, the hallmark of a good government has been, according to the Westminster model of government, the independence and impartiality of the civil service. This is still true in a strategic and enabling state, but in addition there would be some stress on the need to develop an effective partnership across the political-administrative interface so that there is coherence between the political and administrative leadership. The coherence of leadership is an important requirement of forms of governance based on strategic management (Pettigrew et al 1992). A further example from the definition of Government Effectiveness might be the focus on the quality of policy formulation and implementation, and, again, this is still true for strategic states but there might be additional requirements to capture the importance of the quality of *strategic* policy making. None of this detracts from the plausibility that the traditional capabilities of good bureaucratic government could have been quite effective in enhancing a government's environmental performance in the 2000 to 2010 period.

However, since the long-term vision statements launched by the Gulf States were incorporating priorities and concerns about the natural environment, it may well be that after 2010 it would become more and more necessary to deploy strategic-state capabilities to deliver "green" growth and that the bureaucratically based capabilities would no longer be enough.

9. Conclusions

Perhaps the Gulf States have begun to move towards a more strategic-state model of public governance, with general government consumption reducing (as a percentage of GDP) and with them facilitating the development of more supportive environments for competitive business sectors. It has been seen that Voice and Accountability had declined for all the Gulf States. Is this a problem for the modernisation of the state – or does it show that the Gulf States are following their own path to modernization which has more in common with countries in the East like Singapore? It might be remarked that a declining performance on Voice and Accountability would be paradoxical if it coincided with a rise in the power of social media, and if this stimulated a culture in which top down command and control became less acceptable (Osborne and Gaebler 1992).

The conclusions that have emerged about the United Arab Emirates' experience of the period from 2000 to 2010 period have provided a puzzling result in terms of the attempts to view developments from a public governance perspective. On one hand it seems plausible to suggest that the highly rated competence of the United Arab Emirates government using the World Bank's Government Effectiveness indicator was the reason that its record on environmental improvement was second to none among the Gulf States. But then how was it possible to conclude this and observe that there had been a massive drop in terms of GDP growth (on per capita basis) after 2004? If the government had scored highly on Government Effectiveness why had it not steered the economy to growth? In this chapter it has been speculated that some of the puzzling results might be reconciled by presuming a set of traditional capabilities used by bureaucratic governments (not meant pejoratively) to bring about improvements in environmental performance and a different set of capabilities, which it was suggested may be strategic-state capabilities, and which might then explain some of the variation of economic performance delivered by the Gulf states in 2000 to 2010.

Finally, it can be asked if in the future the governments which are well-endowed with strategic-state capabilities will be better placed to deliver higher performance on the natural environment? If so, it can be suggested that this will require that environmental goals stay a top priority in the long-term visions for National development, because strategic and enabling states are believed to achieve greater effectiveness through selectivity and focus, and so higher performance would most likely occur where they remained a key focus of government.

References

Al Diwan Al Ameri (2014) http://www.da.gov.kw/eng/festival/vision_his_highness.php; (accessed 27 January 2014)

Bahgat, G. (1998) The Gulf monarchies: economic and political challenges at the end of the century, The Journal of Social, Political, and Economic Studies, Summer, Volume 23, number 2, pp147-175.

Bahry, L. (1997)The opposition in Bahrain: a belwether for the Gulf?Middle East Policy; May 1997, Volume 5, issue 2, pages 42-57.

Bahry, L. (1999) Elections in Qatar: a window of democracy opens in the gulf, Middle East Policy, volume six, number four, June 1929, pp 118 -127.

Emerson, J.W., A. Hsu, M.A. Levy, A. de Sherbinin, V. Mara, D.C. Esty, and M.Jaiteh. 2012. 2012 Environmental Performance Index and Pilot TrendEnvironmental Performance Index. New Haven: Yale Center for EnvironmentalLaw and Policy.

Harris, J. (2009), Statist Globalization in China, Russia and the Gulf States, Science and Society; Volume 73, Number 1, pages 6-33.

Hertog,S (2007) The GCC and Arab Economic Integration: A New Paradigm, Middle East Policy, Volume 14, Issue 1, pages 52-68.

Hertog, S. (2010) The Sociology of the Gulf Rentier Systems: Societies of Intermediaries, in Comparative Studies in Society and History, Volume 52, Number 2: pages 282-318.

Hvidt, M. (2011) Economic and Institutional Reforms in the Arab Gulf Countries, Middle East Journal, volume 65, number 1, winter 2011, pages 85-102.

Khondker, H.H. (2011) Many Roads to Modernization in the Middle East, Society, volume 48, pages 304-306.

Lahn, G. and Stevens, P. (2011) Burning Oil to Keep Cool: The Hidden Energy Crisis in Saudi Arabia. London: Chatham House.

Moore, P.W. (2002), Rentier Fiscal Crisis and Regime Stability: Business-State Relations in the Gulf, Studies in Comparative International Development, Spring 2002, Volume 37, Number 1, pages 34 - 56.

Nutt, P.C. and Backoff, R.W. (1992) Strategic Management of Public and Third Sector Organizations, San Francisco: Jossey-Bass.

Osborne, D. and Gaebler, T. (1992) Reinventing government: how the entrepreneurial spirit is transforming the public sector, Reading, Massachusetts: Addison Wesley publishing company.
Pettigrew, A., Ferlie, E., and McKee, L. (1992) Shaping Strategic Change, London: Sage.
The International Bank for Reconstruction and Development/The World Bank (1997), The World Development Report 1997: The State in a Changing World. New York: Oxford University Press.

4

Addressing Sustainable Development Challenges in the Gulf Cooperation Council: Imperatives of, and a Framework for Environmental Governance

Jerry Kolo

1. Introduction

There is a robust body of literature that analyzes major environmental challenges facing the Arab Gulf countries that constitute the Gulf Cooperation Council (GCC), in their individual and collective quests for sustainable development. From scholarly analyses of specific environmental challenges, such as the copious research on desertification by Kannan (2012), to professional analyses of general environmental issues (Tolba and Saab, 2009; Tolba and Saab, 2008;), the overwhelming common denominator and consensus in the literature is that the GCC faces both natural and anthropogenic environmental challenges and risks that call for decisive, concerted and in some cases urgent action by all stakeholders in and beyond the region. In terse comments on the challenges, Elouafi (2013), director general at the International Centre for Biosaline Agriculture in Dubai, stated that they "... are very alarming and all of them have been scientifically proven." Instructively, Kannan (2012, p59) noted that, the severity or urgency of the overall environmental challenges in the GCC:

> "... is not because of the interaction between man-made and natural environmental problems but because of the magnitude and speed with which this interaction takes place. The unprecedented levels of industrialization, inefficient use of limited resources, unplanned urbanization, large-scale consumption, higher population growth and lack of regulatory mechanisms produced a critical environmental situation in the GCC countries."

In their general responses to their environmental challenges, the GCC region has demonstrated strong political will by endorsing numerous international agreements and conventions on sustainable development and environmental management; creating regional and national environmental management outfits; allocating or committing public funds to and for environmental causes and initiatives; and continuing to take various other measures to achieve sustainable development. Commenting on these strides, Raouf (2011) noted that "there has been great progress in formulating and executing environmental policies in the GCC," adding that, "in fact, the 1992 Rio Conference accelerated the setting up and strengthening of environment ministries and authorities in GCC states, the adoption of national action plans and strategies, mobilization of financial resources and developing environmental policies." What is clear in the literature, unfortunately, is that the GCC's environmental strides are hampered by a plethora of challenges, such as the sporadic "level of participation ... in international" initiatives, data scarcity and weak legislation (UNEP, 1997); the gap between commitments and action (Wingquist et al., 2012, p12); weak legislation, inadequate financing, oversubsidized water and energy services and public ignorance (Robertson, 2013). Added to these challenges is what this chapter itself deems to be disjointed sustainability initiatives in the GCC.

This chapter posits that the aforementioned challenges would be best addressed through the strategy of environmental governance (EG). Put simply, EG remains the elusive factor, hence the foremost weakness, in the GCC's sustainable development initiatives and endeavors. In perhaps one of the most concise and scintillating analyses of environmental challenges in the GCC, Raouf (2011) noted that "environmental governance in general still lacks many dimensions", and concluded pointedly that "the way to overcome all the ... environmental challenges is good environmental governance". Concurring with this viewpoint, this chapter presents and reviews, in the next section, the concept of EG, deciphers the imperatives of EG, and, based on the imperatives, suggests how the GCC can scale up on, and improve, EG, hence sustainable development, in the region.

2. An Overview of the Concept and Criteria of Environmental Governance (EG)

An overview of the concept and criteria of EG is appropriate at this juncture, first to buttress the contention in this paper that the 'weakest link' in the GCC's sustainability efforts is the lack of a coherent EG system, and, second, to provide the context for the proposal of an enabling framework to scale up EG in the region.

Wingqvist et al (2012, p14) stated that "environmental governance is a specific form of the broader 'governance', and refers to processes and institutions through which societies make decisions that affect the environment". They added that "environmental governance is primarily about how to reach environmental goals, such as conservation and sustainable development, and how decisions are made". In its definition, the World Resources Institute - WRI (2003:viii) stated that EG is "the process and institutions we use to make decisions about the environment", adding that it is about questions concerning "how we make environmental decisions and who makes them" (ibid p6). In this conceptualization, EG involves much more than the work of governments. "It relates to decision-makers at all levels—government managers and ministers, business people, property owners, farmers, and consumers. In short, it deals with who is responsible, how they wield their power, and how they are held accountable" (ibid p6). In line with the WRI, Wingqvist et al. (2012 p15) posited that "environmental governance touches virtually all different aspects of the public sector, from setting the rules of the game, to prioritizing environmental measures and allocating resources. ... It involves multiple actors and is inherently complex and cross-cutting". In another definition, Fakier et al. (2005) stated that "environmental governance refers to the processes of decision-making involved in the control and management of the environment and natural resources. It is also about the manner in which decisions are made". They added that "principles such as inclusivity, representivity, accountability, efficiency and effectiveness, as well as social equity and justice, form the foundation of good governance".

Akin to, and interchangeable with, the concept of EG, is that of Global EG (GEG), which is predicated on the fact that "many of the environmental challenges the world is facing are trans-boundary and must be addressed through joint actions. The international environmental governance system provides an important foundation for addressing these types of common environmental challenges, and the last decades have witnessed a rapid development of the international system of environmental governance" (Wingqvist et al, 2012:20). Najam et al. (2006 p3) observed that "global environmental governance can be seen as the organizations, policy instruments, financing mechanisms, rules, procedures and norms, which regulate the process of global environmental protection". In another definition, Kannan (2012 p28) stated that "global environmental governance is the establishment and operation of a set of rules of conduct that define practice, assign roles and guide interaction so as to enable State and non-State actors to grapple with collective environmental problems across State boundaries". This view of GEG is shared by Wingqvist et al. (2012:10), who noted that "environmental governance is cross-cutting, relates to international, national, and sub-national levels, and involves many actors".

The summation in this chapter is that 'governance' provides the conceptual pedigree for EG, hence the intrinsic relationship between the two. As Wingqvist et al. (2012:14) rightly noted, EG should be examined in the broader context of 'governance', which is a concept "generally used to describe how power and authority are exercised and distributed, how decisions are made, and to what extent citizens are able to participate in decision-making processes". In this sense, they elaborated further that:

> "Environmental policy design is embedded in a political context with multiple actors and interests. In many cases measures that strengthen important human rights principles, such as the rule of law, transparency and public participation, may be equally or more important than specific environmental policies or projects in order to improve environmental management. It is increasingly recognized that technical solutions to environmental problems are not sufficient to obtain sustainable development. Instead, there is a growing attention to the importance of governance to manage the wide range of environmental challenges and impacts" (Wingqvist et al., 2012 p8).

The view in this chapter is that EG, and by extension GEG, provides an excellent canopy for trans-border or multijurisdictional cooperation in sharing knowledge, skills, experiences, in short 'best practices', for addressing local, national, regional and global environmental issues cost-effectively. GEG is anchored by the global community's realization and acknowledgment that the world is better off addressing unique and common environmental challenges through:

a. Shared environmental values and vision for global sustainability through full respect for the rights of societies to grow and develop while respecting and safeguarding the natural environment or life-support systems (LSS).
b. Collective responsibility to current and future generations by using natural resources prudently to provide dignified standards of living.
c. Partnerships and collaborations, devoid of ideology, to leverage the world's limited resources to address the world's vast, complex and intertwined environmental challenges.
d. Full accountability to specific and broad constituencies and stakeholders on environmental matters under the principle of accountability.
e. Equitable access to environmental decision processes and information under the transparency principle.

A variant of the criteria listed above was included in a government white paper in South Africa, referenced by Fakier et al. (2005 p5). According to that document, the "pointers as to what constitutes good environmental governance" are:

- Governance should be responsible and accountable;
- Regulations should be enforced;
- Integrating mechanisms and structures that facilitate participation should be established;

- There needs to be inter-ministerial and inter-departmental co-ordination;
- The institutional responsibilities for regulating environmental impacts and promoting resource exploitation should be separated;
- People should have access to information; and,
- There needs to be institutional and community capacity building.

Fakier et al. (ibid p5) suggested that "an eighth element that covers the integration of environmental issues into other sectors (mainstreaming) can be added".

In another version of the above-listed criteria, Raouf (2011) opined that:

> To achieve good environmental governance, there is a need for community participation in drafting policies. Besides, proper allocation of authority needs to be enhanced and environmental institutions must be empowered. Strengthening the role of various stakeholders (NGOs, the private sector, local communities, etc.) would improve execution, monitoring, reporting and contribute towards achieving the collective goals as well as increasing the cooperation at national and regional levels. This would also lead to better implementation of environmental policies".

3. Imperatives of EG

This chapter uses the various criteria listed above as the basis for delimiting the following imperatives of EG. In turn, the imperatives are fused into the framework proposed for EG in the GCC. The imperatives of EG are:

1. Policy making – to capture the visions and aspirations of the GCC for sustainable development, set priorities accordingly, and ensure accountability and transparency of the EG system.
2. Legislation – to cast in legally binding terms the terms and conditions, processes and mechanisms, and penalties and incentives for managing environmental resources in the region.
3. Budget – to adequately fund all aspects and activities of sustainable development and environmental protection in the GCC.
4. Administration and management – appropriate structures and institutions with the capacity required to effectively lead and coordinate the implementation of sustainable development initiatives in the GCC.
5. Intergovernmental coordination and public-private partnerships.

6. Research, education and training – to build and nurture technical and professional capacity, and to enhance citizen awareness.
7. Community outreach and citizen mobilization – to energize civic responsibility and citizenship, civic engagement, citizen enlightenment and advocacy.

4. An EG Framework for the GCC

The framework proposed in this chapter is to enable the GCC to scale up its EG system. The idea of scaling up is predicated on the fact that EG exists in the GCC but in a very rudimentary stage. As Raouf (2011) clearly noted, "in general, GCC states have made considerable progress in environmental governance. However, the trends show the need to make use of additional policy tools, and engage the various stakeholders in the governance process". This chapter submits that the urgency to scale up on EG is indubitable, in light of the quantum pace of growth and development, hence, environmental pressures and challenges, in the region. However, most profound, and most pertinent to the thrust of this chapter, is the finding that environmental issues and requirements are still not amply integrated in the long-term development planning of the GCC, nor are they adequately addressed at the appropriate levels of policy and decision-making. This finding points directly to a critical but feeble dimension of EG, which is the subject of this chapter. Significant strides, albeit structurally and functionally disjointed, have been made and continue to be made since the referenced study. Yet, evidence on the ground suggests that the GCC has a long way to go for EG, one of the most critical requirements for sustainable development, to gain traction and become effective standard practice in the region.

Using the seven EG imperatives delineated earlier, details of the EG framework proposed for the GCC are shown in Table 1 below. Most unique about this framework are the following.

1. The framework would decant planning and managing what this chapter calls the GCC's Critical Resources of Regional Priority (CRRP s) from the current political processes and machines that seem to condone, perpetuate and perpetrate waste, inequity and ad hoc approaches to environmental protection. The planning and management functions would be assigned or entrusted to one independent, technical, professional and apolitical agency that would serve the whole region. For the proposed EG framework, this chapter refers to the agency as the GCC Regional Sustainability Planning Agency (RSPA).
2. Through a regional referendum of GCC policy makers, GCC member States would pass legislation authorizing the establishment of a RSPA as the professional public agency responsible for planning and managing designated CRRPs in the region.

Legislation would establish and determine the size of the governing board. CRRP would be identified and assigned to the RSPA for planning and management. A Regional Sustainability Board (RSB) would be established as the policy making arm of the Agency, and the method of selecting or appointing members from all the stakeholder sectors of society to serve on the board would be specified in the legislation.

3. RSPA would operate independently from all member States' governing bodies and bureaucracies, but in full and cordial partnership with them.
4. Premium will be placed on technical and professional competency in appointing members to serve on the governing board (RSB), and to work in the RSPA.
5. The budget to run the Agency would come from member State dues, and from fees charged for professional and technical services performed or provided by the Agency, such as fees from processing development permits for projects of regional impact (PRI), and fees charged to provide technical assistance and training for GCC municipalities on sustainable development and environmental protection skill and methods. The board would lobby policy makers in member States for dedicated revenue sources, such as tourism bed tax, or a surcharge on paid parking across the region.
6. CRRPs would be allocated to consumers across the GCC, based on established scientific data and standards, full-cost recovery (paid by consumers or subsidized by member States for their citizens), and strict adherence to sustainability principles and standards.

The roles and responsibilities suggested for the main stakeholder actors in the EG framework and process are described in Table 1.

5. Conclusion

From macro global environmental challenges such as climate change to micro local challenges such as diminishing and congested open spaces, the world in general, and the GCC in particular face environmental challenges, many of which this chapter believes are nearing the what may be termed the 'tipping point' (Gladwell, 2002) and almost crossing over into the crisis stage. In a sobering reflection on UAE's water scarcity problem, for example, Ahmed Al Mansouri (2013), a member of the UAE Federal National Council, did not mince words when he stated that "now, we're talking about risk management but in a few years we'll be talking about crisis management. We can't live in the comfort zone".

In spite of the severity and urgency of the GCC's environmental challenges, the ever looming risks and threats of what March and Grossa (1996 p7) called the four dilemmas of environmental sustainability cannot and must not be ignored or discounted.

Table 1 Roles and Responsibilities of Main EG Stakeholder Actors

EG Imperatives	Roles and Responsibilities of Main EG Stakeholder Actors		
	GCC/Regional	States/National	Non-State Actors
1. Policy making	a. Establish Regional Sustainability Board (RSB), based on proportional representation by main stakeholder sectors, to serve as governing board, with executive powers. b. Designate and oversee planning and management of Critical Resources of Regional Priority (CRRP) c. Establish, coordinate and facilitate regional and international collaboration and partnerships. f. Capture the visions and aspirations of the GCC for sustainable development, set priorities and ensure accountability and transparency of the EG system.	a. Designate/appoint members to RSB b. Coordinate State plans with regional plans (principles of integration and subsidiarity) c. Serve as bridge between RSB and sub-national public entities.	a. Mobilize non-State actors to lobby and advocate for constituency environmental interests in the GCC and globally. b. Nominate competent representatives to serve on RSB.
2. Legislation	a. Make EG policies, laws and regulations for sustainable development and environmental protection in the GCC. b. Adopt legislation mandating public hearings on all RSPA projects at designated locations across the GCC.	a. Provide adequate resources and political support for RSPA to function independently	a. See 1a above.
3. Budget	a. Oversee budget and revenue generation for RSPA, and ensure public accountability and transparency.	a. Pay membership dues to RSPA for operating and capital budgets. b. Support revenue generation plans of RSPA, e.g., dedicated revenue source	a. Take steps to hold RSPA, via RSB, responsible and accountable to the public.
4. Administration and management	a. Formulate plans for CRRPs, submit to RSB for approval and funding b. Coordinate and/or implement, monitor and evaluate plans approved by RSB. c. Review, amend, approve and monitor projects of regional impact (PRI) d. Facilitate vertical and horizontal integration of RSPA plans and activities across GCC.	a. Submit and get approvals for projects of regional impact (PRI) to and from RSPA.	a. Participate in RSPA activities as necessary.

	e. Conduct or sponsor research, and develop, maintain and share robust GCC database and maps on GCC environment. f. Provide environmental training and technical assistance for capacity building across GCC. g. Design, disseminate and promote grassroots outreach materials on GCC environment. Facilitate outreach across GCC. h. Collaborate with grassroots organizations and non-State actors to organize grassroots activities to promote environmental education, awareness and model practices.		
5. Intergovernmental coordination and public-private partnerships.	See 1c and 4d above.	a. Serve as bridge between RSPA and sub-national environmental public agencies.	a. Support and facilitate partnerships and coalitions by RSPA for sustainable development
6. Research, education and training – to build and nurture technical and professional capacity, and to enhance citizen awareness.	See 4e-g above	a. Collaborate with RSPA to promote environmental research, education and training in the GCC	a. Conduct research and produce reports and data to inform environmental policies and plans by RSPA and across the GCC.
7. Community outreach and citizen mobilization – to energize civic responsibility and citizenship, civic engagement, citizen enlightenment and advocacy.	See 4h above.	a. Collaborate with RSPA to promote community outreach and grassroots participation in environmental protection.	a. Organize and sponsor forums, conferences and community outreach to raise environmental awareness and activism at the grassroots.

Source: Framework constructed by the author, 2014

In sum, the dilemmas are population explosion, consumerism, the 'exploitative' capacity of technology and the unending debate on what sustainability really means. The view in this chapter is that these issues and dilemmas are priority questions, concerns and premier agenda items of responsive, accountable, transparent and participatory governance, hence the conviction and proposal that EG is the best strategy for sustainable developmentand environmental protection in the GCC and, indeed, the world community. The EG framework proposed in this chapter incorporates what the paper delineated as the imperatives of EG.

The discourse and proposed EG framework in this chapter focus on the GCC as a single environmental planning entity because of the broad geo-spatial, environmental, religious, economic, cultural and even political similarities of most or all the countries that constitute the GCC. In this sense, concerted efforts to address the environmental challenges facing the region should be in the best interest of the member States for some basic advantages, including:

i. Cost-sharing and financial savings in implementing environmental policies, plans and projects.
ii. Leveraging and complementing domestic and foreign capacity for environmental planning, management and development.
iii. Sharing and exchanging technical knowledge, skills, experience and best-practice models, rather than 'reinventing' the wheel in individual countries.
iv. Encouraging coordinated and forestalling disjointed responses to regional environmental challenges.
v. Minimizing overreliance on foreign technical and professional environmental expertise.
vi. Reinforcing economic, technological, political and strategic bonds and partnerships in the region.

Daring to propose an EG framework for the GCC as this chapter is based on the conviction that political will, technology and financial resources in the GCC are necessary but insufficient to mitigate the constant environmental damages inflicted on the region's natural environment by growth and development, and to safeguard the environment for the fair and equitable benefits of current and future generations. Governance, and by implication EG, anchored on the pillars or principles of citizen engagement, citizen enlightenment, institutional (vertical) and sectoral (horizontal) integration, transparency, accountability and social justice, must become the overarching framework under which political will, technology and financial resources are deployed smartly, strategically, ethically and responsibly to protect the environment, pursue sustainable growth and development, and discharge the world's moral obligation to future generations.

References

Al Mansouri, A. (2013) 'Water scarcity will be at 'alarming levels' by 2025, GCC warned.' Interview by Caline Malek. *The National* [Online] 16th September 2013. Available from: http://www.thenational.ae/uae/water-scarcity-will-be-at-alarming-levels-by-2025-gcc-warned#ixzz2f8g4Tr00 [Accessed: 22nd March 2014].

Elouafi, I. 92013) 'Water scarcity will be at 'alarming levels' by 2025, GCC warned.' Interview by Caline Malek. *The National* [Online] 16th September 2013. Available from: http://www.thenational.ae/uae/water-scarcity-will-be-at-alarming-levels-by-2025-gcc-warned#ixzz2f8g4Tr00 [Accessed: 22nd March 2014].

Fakier, S., Stephens, A., Tholin, J. and Kapelus, P. (2005) Environmental Governance: Background Research Paper produced for the South Africa Environment Outlook report on behalf of the Department of Environmental Affairs and Tourism. Available from: http://s3.amazonaws.com/zanran_storage/soer.deat.gov.za/ContentPages/42728136.pdf [Accessed: 21st March 2014]

Gladwell, M. (2000) *The Tipping Point: How Little Things Can Make a Big Difference*. New York: Little, Brown and Company.

Kannan, A. (2012) *Global Environmental Governance and Desertification: A Study of Gulf Cooperation Council Countries*. New Delhi, India: Concept Publishing Co. Pvt. Ltd.

Marsh, W. M. and Grossa, Jr., J. M. (1996) *Environmental Geography: Science, Land Use, and Earth Systems*. New York, NY: John Wiley & Sons, Inc.

Najam, A., Papa M. and Taiyab, N. (2005) *Global Environmental Governance: A Reform Agenda*. Winnipeg, Manitoba: International Institute for Sustainable Development. Available from: http://www.iisd.org [Accessed: 25th February 2014].

Raouf, M. A. (2011) Environmental governance in GCC is key. *Gulf News*.[Online] 23rd September. Available from: http://gulfnews.com/opinions/columnists/environmental-governance-in-gcc-is-key-1.871590 [Accessed: 23rd February 2014].

Robertson, A. (2013) Middle East sustainability must not be lip service. *Gulf News*. [Online] 27th June. Available from: http://gulfnews.com/business/property/uae/middle-east-sustainability-must-not-be-lip-service-1.1202579 [Accessed: 22nd February 2014].

Tolba, M. K. and Saab, N. W. (eds.) (2008) *Arab Environment: Future Challenges*. Beirut, Lebanon: Arab Forum for Environment and Development (AFED). Available from: http://www.afedonline.org [Accessed:23rd February 2014].

Tolba, M. K. and Saab, N. W. (eds.) (2009) *Arab Environment: Climate Change – Impact of Climate Change on Arab Countries*. Beirut, Lebanon: Arab Forum for Environment and Development (AFED). Available from: http://www.afedonline.org [Accessed: 23rd February 2014].

UNEP- United Nations Environment Program (1997) *Global Environmental Outlook – 1: Global State of the Environment Report*. Nairobi, Kenya. Available from: http://www.unep.org/geo/geo1/ch/ch3_29.htm [Accessible: 22nd February 2014].

Wingqvist, G. Ö., Drakenberg, O., Slunge, D., Sjöstedt, M. and Ekbom, A. (2012) *The role of governance for improved environmental outcomes: Perspectives for developing countries and countries in transition*. Stockholm, Sweden: The Swedish Environmental Protection Agency. Available from: www.naturvardsverket.se/publikationer [Accessed: 23rd February 2014].

World Resources Institute (2003) *World Resources 2002-2004- Decisions for the Earth: Balance, Voice, and Power*. Washington, DC: World Resources Institute. July 2003
Available from: http://www.wri.org/sites/default/files/pdf/wr2002_fullreport.pdf [Accessed: 22nd February 2014].

5

Contribution of the Private Sector towards Inclusive and Sustainable Growth in the Gulf Region

Yousuf Hamad Al-Balushi

1. Introduction

This chapter seeks to highlight some of the critical issues considered to be important for the future development of the private sector, positioning it as an engine of inclusive growth and prosperity for the gulf region.[1] To do so, it will be divided into four themes. The first theme will provide an introduction to the global environment, recognizing that the GCC countries are highly integrated into the world economy, but that realizing the potential gains from such integration is difficult without involvement of a vibrant and competitive and self-reliant private sector (Al-Barwani, 2008). The chapter also discusses a variety of issues considered to be important for the continuation of the private sector led growth in the Gulf region, which at present has a minimal role in the economy in a hydrocarbon dependent sector. The second theme analyses the challenges associated with the private sector for sustainable growth, specifically the labour market structure and national identity in the sector, noting the young and rapidly growing national labour force, the heavy reliance on expatriate labour in the private sector and a large skill gap between job seekers and the market requirements. Moreover, it discusses the financial sector and the need to expand credits to the productive sector, promotion of saving behaviour, and why economic diversification and redistribution of rent revenues have so far produced a national bourgeoisie rather than local entrepreneurs. The third theme will look at the Institutional framework needed to support and enhance the transformation, the role of state-owned enterprises in leading the new trend and opening the gate for firms to grow and the role of foreign direct investment (FDI) in private sector development.[2] In addition to addressing the above-mentioned themes, this chapter also provides actionable policy recommendations designed to address some of the constraints that inhibit the private sector from growing. These include improving the business environment with special attention to SMEs, improving access to finance, enhancing management and business development capacity through training, linkages and networks and creating an institutional mechanism to support development.

2. Private Sector Theoretical Analysis

The ultimate objective for any given economy is to achieve sustainable development, based on economic diversification and growth. One way to achieve this is by diversifying economic outputs; this is mainly done in developed and industrialized countries through exploring different market and different needs (Sitglitz, 2002). In developing countries, the more usual approach is diversifying the economic inputs through diversifying the income sources. Since this chapter focuses on developing countries, particularly GCC, diversification of income sources is the target. As illustrated in figure 1 diversifying the economic income is achieved through improving the performance and competitiveness of the private sector (World Bank, 2009).

Figure 1 **General Framework for Sustainable Development**

FDI
Private Sector
Improve performance and competence of different economic sectors
Diversify Input/ Income Resource
Economic Diversification and Growth
SUSTAINABLE DEVELOPMENT
In Developing Countries
In Developed Countries
Business Environment
Government
Diversify Output
New Market New Product

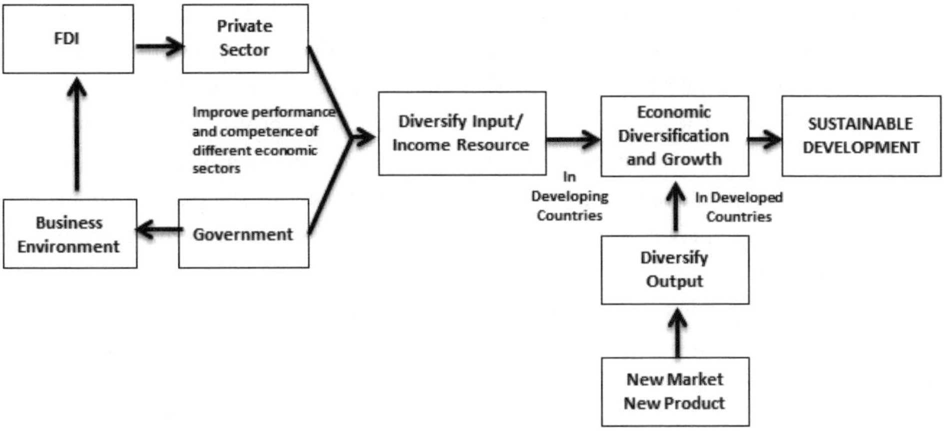

Source: Compiled by the author

Recently, there have been significant changes to traditional concepts of development, due to globalization and the communications revolution. Development planning is no longer dependent on local resources only. Through globalization, GCC countries can use the world's resources and simply manage them into final product. For example, they can utilize funds for a project from one country, manpower from another, materials from a third country, and yet deliver the final project in a totally different country simply by "smart management" of the resources. Globalization has been pushed mainly by business, specifically Foreign Direct Investment (FDI) and Multinational Corporation companies' (MNCs) activities; all countries are competing with each other to attract maximum FDI (Cohen, 2006). FDI has substantially grown worldwide over the last three decades and due to globalization it is the most popular strategy for the enhancement of the private sector. This is due to a variety of possible channels for spillover benefits that might boost economic performance in the domestic private sector, which include imitation, skill acquisition, competition and exports. These spillover benefits can accrue through different types of business linkages between domestic companies and MNCs such as backward linkages with suppliers, linkages with technology partners and forward linkages with customers (Gorg and Greenaway, 2004). However, these benefits do not accrue automatically and should not be taken for granted, as MNCs will not simply hand over the source of their advantage (Cohen, 2006). The host economy needs to prepare the ground and pre-conditions for successful spillover benefits from FDI companies to domestic ones through well designed and implemented linkages programme and investment map.[3] In fact, integration of the private sector with the global economy needs a competent private sector to serve as the growth engine and a driving force for the development process in this region. The private sector is pivotal in all countries that have grown strongly over long periods. The Commission on Growth and Development found in their 2008 report five common characteristics of countries with high-sustained growth.[4] While many of these characteristics relate to actions that governments need to take, also prominent among these is market allocation of resources, led by the private sector.[5] According to the same report, in most countries, the private sector is the major component of the national income, the major employer and creator of jobs. Over 90 percent of jobs in developed countries are in the private sector. The pace of job growth and the quality of employment in the private sector are thus central to the development (World Bank, 2005).

Literature highlights several potential factors as key determinants of private sector development; these factors can be classified into factors related to government and factors related to the private sector itself. In this part, we will address the Pre-conditional Determent Factors for private sector led growth to materialize.

Figure 2 **Pre-Conditional Determent Factors**

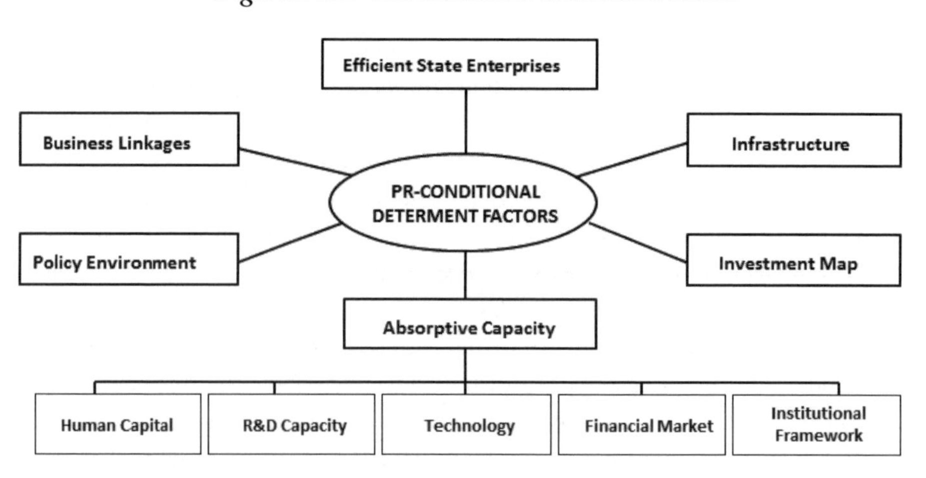

Source: Compiled by the author

Absorptive capacity is an important factor affecting the extent of private sector development (Mishrif, 2010). It is, however, important to distinguish between absorptive capacity at the firm level and at the country level. Absorptive capacity at national level is shown by indicators such as per capita income, trade openness, the level of education of the labour force, the level of development of the financial markets, use of technology efficiency and domestic research and development. At firm level, it is linked to the quality of human capital, and management, highlighting the role of education and management training policies. The importance of R&D in expanding the technology frontier is discussed by Aghion and Howitt (1992) and Grossman & Helpman. (1991). The dominant variable in most studies is the standard of education of the labour force, which is also described as "social capability", a threshold level of human capital and human capital formation (Cohen and Levinthal, 1990). Literature highlights five important elements related to absorptive capacity, namely; Human Capital, Research and Development capacity, Institutional Framework, Financial Market, Policy Environment and Intellectual Property Rights.

Available statistics and official reports confirm that there is a shortage of absorptive capacity in the region, particularly in terms of needed institutions to drive the private sector led growth not in terms of quantity but quality. Also, technical and managerial skills are needed to drive this transformation and private sector competitiveness. The quality of current institutions should be enhanced and new institutions created to build linkages and scale up success stories e.g. SMEs support education and training.

Globalization has played a key role in inducing economic growth around the world. Opening up the economy is expected to have a direct impact on the local economy by creating demand for a broad range of industrial and consumer goods resulting in more jobs, enhancing the pace of economic growth and creating more opportunities for entrepreneurs. The GCC has become more liberal in trading with increasing integration in the world economy, since they formed a Free Trade Area in the early 1980s, eliminating all tariffs on local products. In 2002 its members agreed to form a customs union with a unified external tariff ranging between 5% to 10% (Ramady, 2012). This trend of liberalization is evident from a number of links such as the high ratio of openness of the GCC region amplified by exports of crude oil,[6] and the sharp increase in FDI, as well as expatriates contributing to more than 75% of total labour force in this region. In order to manage integration with the global economy and enable the private sector to grow, the region needs to improve its institutional framework[7] and identify forward and backward linkages with the world economy where they exist. A number of studies have indicated that the positive impact of FDIs on growth depends on local conditions and absorption capacity. In order to achieve sustainable growth, there needs to be significant interactions between FDIs and other economic sectors. In promoting FDI careful monitoring is needed to avoid negative impact on long-term development as well as environment drawback. Foreign investors should realize that investment in any one GCC country would give access to all Gulf countries. In order to encourage foreign investment to make positive contributions to the development of the private sector and economic development, some additional measures may need to be taken. Four criteria are recommended for choosing foreign investors for GCC. The first criterion is to make productive use of our natural resources. The second criterion is that of providing some added value to local society in the form of developing our human resources. A third criterion is reduction of our import cost. Fourth is the creation of a source of export revenue. Recognizing the small local market and increase in globalization, it is essential to focus on creating global and regional integration to boost productivity and growth. As the world economy becomes more open and tightly integrated, sustained growth becomes feasible and GCC countries can benefit in two ways: first, by importing ideas, technology and knowhow from the rest of the world; second, by exploiting global demand, which provides a deep elastic market for the goods. Economic liberalization should ensure fair and open competition in the market, thus creating opportunities for a more efficient allocation of resources and support for private sector investments and growth.[8]

Figure 3 **Private Sector Led Growth Model**

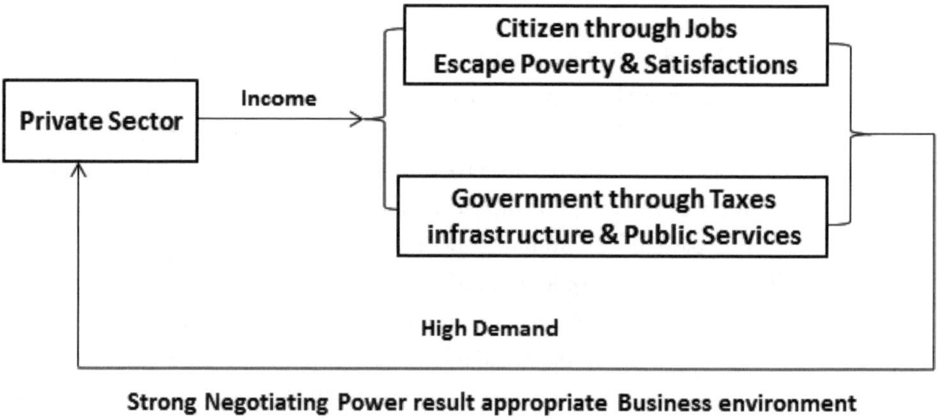

Private Sector
Income
Citizen through Jobs
Escape Poverty & Satisfactions
Government through Taxes
infrastructure & Public Services
High Demand
Strong Negotiating Power result appropriate Business environment

Source: Compiled by the author

As illustrated in figure 3, the private sector is the main driver of the development process in advanced and a number of emerging economies, while in most oil-based economies such as the GCC countries, primary income accrues mainly to government as the owner of the natural resources (oil and gas), which has helped the government play a major economic role as investor, provider of social services and infrastructure and in some production activities. Statistics indicate weak performance of the private sector during the last four decades. Although there is significant improvement in private sector performance,[9] these countries have yet to reach the target level, even with the great support received from governments through the enactment of laws, infrastructure development, provision of various types of incentives and support to the private sector, commitment to market mechanisms in resource allocation, and encouraging free competition to stimulate innovation (Hertog et al., 2013). GCC governments understand that the state-led development scenario used previously will not sustain long term growth and address challenges and there is an urgent need to move to a new model with a larger role for the private sector.[10] Much has been done but still more remains to be done to align the incentives of profit maximizing in the private sector with the social objectives of shared growth and job creation. As indicated in various World Bank reports, the private sector in developing countries, including the GCC, faces many constraints in finance, infrastructure, employee skills, and the investment climate.

3. Overview of the GCC Economies

The GCC regional macroeconomic performance in 2013 continued to remain robust; and overall, the Gulf region economy has performed well in the last 40 years,[11] characterized by strong GDP growth,[12] significant creation of job opportunities, and large surpluses in the fiscal and balance of payment positions (GCC, 2012). The economies of the Gulf region have attracted increasing attention over recent years, especially with the increase in oil prices (Ramady, 2012). These economies have also become more important as global investors and trade partners that play a crucial role in global energy markets. Furthermore, together with other major oil-exporting countries, they have become part of the international policy debate on global imbalances. On the global level, all GCC countries are actively trying to integrate into the world economy and making a move toward the market system[13] Nevertheless, the region faces numerous constraints, including the absence of a dynamic and vibrant private sector. The GCC countries share a number of specific structural economic features. Key common features are a high dependency on hydrocarbons as expressed by share in oil and gas revenues from export and the share of the hydrocarbon sector in GDP, and a young and rapidly growing national labour force, heavy reliance on expatriate labour in the private sector and a low degree of self-sufficiency in almost all requirements except in hydrocarbons. These states also have a high degree of openness and stable economic framework characterized by overall low inflation, stable growth and stable currency. That said, these features also pose common structural policy challenges to GCC economies, notably economic diversification and education resulting in a lack of employment for the locals while creating huge demand for the foreign labour and sustaining the same level of economic growth. Although the private sector in the region is significantly growing and expanding, it is important to distinguish that this growth is driven by a number of government support policies which made capital, energy and infrastructure available at low or sometimes no cost.[14] Cheap utility services including gas, electricity and water and loans have enabled both private and public companies to quickly make profit and expand. However, this provides little or no incentive for upgrading productivity, investing in technology and engaging in research and development (Hertog, 2012).

As evident in figure 4, the private sector in the GCC is characterized mainly by businesses that are personal or family-owned, offering low-wage and low-skill employment, resulting in the fragmentation of this sector to small enterprises that lack the financial means and skills needed for doing business on a large scale. In fact, private sector enterprises favour the hiring of expatriates who accept lower wages, require less training and are subject to more flexible labour market regulations, compared to the local workforce.

Figure 4 **Private Sector Characteristics in GCC**

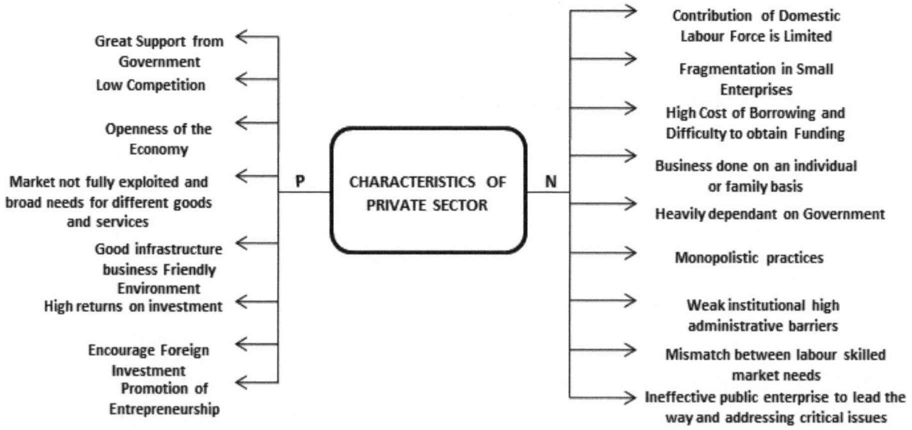

Source: Compiled by author

Developing a strong private sector that is able to ensure a growth pattern that generates productive employment opportunities for nationals and becomes the leading sector in the national economy is a top priority in all vision and development plans. However, despite the many successes, these countries still have not been able to realize this objective. This part will dwell a little on the role and the contribution of the private sector in the main economic areas, followed by challenges that prevent high performance.

4. Role and Contribution of the Private Sector towards Employment and Labour Market

There are a number of features that could explain labour market structure and challenges in the region. These include first, the young and rapidly growing national labour force and the heavy reliance on expatriate labour in the private sector, as over 65% of the population is under the age of 25 years. Additionally, the population growth rate is 3.5% annually. On the other hand, private sector enterprises make little contribution to national employment, on average accounting for 20% of the whole private sector labour market, compared with 80% for expatriates. Second, education policies result in a large skill gap between job seekers and the market requirements. According to the World Bank (2008), access to education in the region has improved. However, quality is deficient with schools unable to help students develop basic and technical skills, creating a mismatch in skill between job seekers and the labour market.[15] It is also important to recognize the huge financial cost associated with remittances, which according to the GCC Secretariat accounted for $70 billion in 2012, in addition to increased pressures on services such as education and health and utilities including electricity and water, and more recently, increased demand from international organizations concerned with labour rights including nationality. Third, government-spending packages crowd out the private sector[16] resulting in high dependence on employment of nationals on the government (Al Qudsi, 2008). In fact, in the government sector the labour force on average is 80% nationals and 20% expatriates. For example recent statistics show that the public sector is the main source of job creation for nationals, in 2012 employing 80% of nationals in Saudi Arabia,[17] 85% in Oman, and 93% in Kuwait. In most GCC states, for political reasons and to ensure national stability, governments resort to creating unproductive jobs in the police, defence, national security, and civil services. Young nationals expect to get a job in the government sector sooner or later; because it works as security enabling them to get large sums of loans from banks for automobiles, marriage, and homes. Overall, this scenario does not motivate them to work hard, pursue a high quality of higher education and accept the hard path of private sector employment. When asked in a recent Silatech-Gallup poll whether they would prefer working for the public or private sector or to start a business, between 60 to 80% of the Gulf youth favoured the public sector (Silatech and Gallup, 2009). However, according to the World Bank (2008), governments will not be able to create these jobs in the public sector, nor will state-owned enterprises in a sustainable manner. The jobs will have to come from the private sector. According to Hartog (2012), the private sector in the Gulf region has created an enormous number of jobs. However, the majority of these are held by expatriates as they are low-paid and low-productivity.[18]

Challenges in the labour market have multiple dimensions. First, there is a weak link between the job market and education and training programme, resulting in poorly prepared new entrants to the job market. Second, there is poor quality of basic education, which is not producing students who can adapt well to the needs of a specific job. Additionally, availability of low-cost expatriate labour has driven wages in many sectors well below what nationals would willingly accept. Governments need to give more attention to the balance between political stability, huge fiscal cost and long-term commitment in recruiting the increasing numbers of job seekers. It is also important to promote entrepreneurship and directing nationals to take the business path instead of serving in the government. At the same time one has to strike a balance between rules that enforce employment of nationals on the one hand and allow the private sector to be competitive through addressing the challenges faced by education. In addition, to meeting labour market demand, it is necessary to re-evaluate existing manpower institutions and encourage a new institution of high technical and managerial proficiency.

5. Role and Contribution of the Private Sector in Economic Growth

The GCC countries future vision for private sectors encompasses three broad goals. First, that the private sector should be characterized by increased efficiency and integration with the global economy. Second, that the private sector should be the main source of economic activity.[19] Third, that the private sector should be the main source of employment for nationals.

Figure 5 GCC, State Led Growth Model

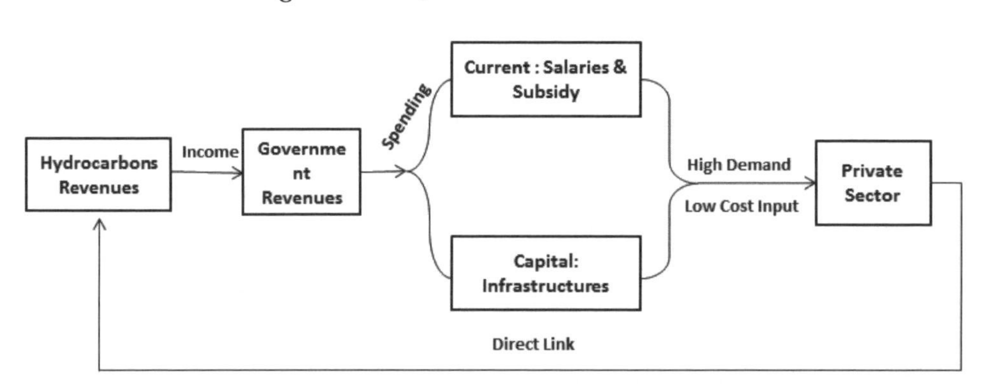

Source: Compiled by author

Despite the visionary goals and improved performance in recent years it has still not reached the target level (Ramady, 2012). As illustrated in figure 5, the private sector makes little contribution to the economy in terms of value addition. This is because the public sector plays the dominant role in the hydrocarbon sector, contributing through oil and gas. The GDP growth trend over the years has been largely driven by crude oil prices in the international market. The effects of crude oil prices are also manifested through the inflation rate, employment, investment and saving.

6. Role and Contribution towards Economic Diversification

Generally, the degree of economic diversification is reflected by sectoral composition. The GCC sectoral composition as shares of the nominal GDP shows mixed results influenced largely by trends in oil prices. Higher oil prices often skew petroleum activities share of GDP while minimizing the share of non-petroleum activities and vice versa (Al-Barwani, 2008). Since the beginning of the GCC modern economy, diversification of the economy has taken priority in all plans (Hammad, and Washington, 1986). A diversification strategy that can cope with the risks of depletion of oil and gas reserves should emphasize the development of sources of growth and exports that are not dependent on hydrocarbon input, or can still compete internationally when lower costs of domestic inputs are no longer available (Fasano and Iqbal, 2003). In addition to public investment, structural policy reforms are likely to be crucial to allow private domestic and foreign investment to flourish and naturally develop areas of comparative advantage. Thus, policy makers are faced with the challenges of how to transform the structure of the national economy in order to reduce dependence on oil (Hvidt, 2013). The lack of major alternative natural resources, lack of skilled native manpower and high defense expenditures contribute to increasing such challenges. Governments have taken important steps to develop the manufacturing sector, enhance the quality of the labour force and increase job opportunities, improve the country's infrastructure, make the business environment more attractive and stimulate FDI. Nevertheless, as statistics indicate, although the share of non-oil (tradable) is increasing, its size remains relatively small when compared to the shares of oil, gas and services sectors.

7. Role and Contribution of the Private Sector towards Capital Formation

It is worth noting that the GCC, with the exception of Kuwait, maintains a fixed peg to the US dollar.[20] In general, across the region, the financial sector has a strategic role but is not treated strategically (GCC, 2012). It is characterized by:

- The government is the dominant saver and investor. This is reflected in the relative share of public capital formation in gross capital formation.
- The various government policies serve to crowd out private savings for example, subsidy and pension systems, that remove any private saving incentive.
- Financial institutions do not channel savings in accordance with the needs of productive investment, which is essential in achieving long-term growth, main business lines consumer loans to households.
- Private investment is not satisfactory. In fact it is weak; national savings are less than domestic savings due to expatriate labour remittances and interest and dividend paid on external liabilities. The private sector's contribution to total investment, although increasing over the years, remains the lowest among developing regions.
- Commercial banks are among the most profitable institutions in the region.
- Due to the limited number of commercial banks and lack of competition, services and policies are not designed according to the market need.
- Banks engage in relatively little long-term lending, project financing is not common in the market.

- The SME-specific segments of banking sector are essentially niche products for the banking system (Hertog, 2010); their main business is with salaried households or medium-to-large corporate clients.[21]
- Public investment is dominant over private investment. The increase in domestic savings has not necessarily contributed to financial sector development, because private savings remain low.[22]
- The low private savings rate is unusual in a country with a high-income level when income and saving have a strong positive relationship. [23]
- Savings incentives are further blunted by a relatively big gap between lending and deposit interest rate
- Finance and leasing companies remained marginal to banks
- Insurance industries are still small compared to banking in the region.
- Public sector investment accounts for an average of 75% of the total investment.
- Monetary policy structure is inadequate.
- Credit distribution among the active sector needs to be incorporated in the new rules of Commercial Banks for future development.

There is a need for increased emphasis on the financial sector as a critical pillar of private sector led growth and it should be much more of a vehicle for channeling the domestic saving to productive sectors instead of consumption. Equally important is to tackle common barriers, which may include access and high cost of credit, the collateral constraint, inappropriateness of lending products, unavailability of long-term loans at competitive rates and others, supporting and promotion of private investment and non-oil exports. Also, developing SMEs is the backbone to developing an active and more efficient private sector. In addition, opening new foreign banks could help in attracting foreign investment.

8. Conclusion and the Way Forward for Private Sector Led Growth

The above discussion shows a clear sign of a performance problem with private sector development and growth in the GCC countries, as a result of some fundamental impediments as everything is connected and matters for private sector growth. Understanding and addressing the existing concerns requires an in-depth coordination between the three-pillars: government, business and society[24] to ensure a win-win situation is achieved.[25] Moreover, private sector development is a multi-sector challenge and will not happen naturally; it needs to be worked at. The slow progress of the GCC private sector could be attributable to the unusual relations between government, business and society and equally important lack of functional institutions to lead private sector transformation. Indeed, it is crucial for private sector led growth to remove conflicts of interest between politicians and business people. Having said that, there is an urgent need to tackle typical barriers facing private sector growth and recognize what determines this growth. This includes dealing with genuine problems in all production factors including capital, labour and institutions, and strengthening absorptive capacity (World Bank, 2009). The culture of doing business in the region must also be considered. In fact, to meet the above challenges is not an easy task; it requires development of multiple institutions associated with finance, labour, regulation, and trade, as well as the accumulation of production, quality and marketing know-how by firms and workers. The dilemma is to create the balance among them. In light of the expected depletion of oil and gas reserves, the region will need to develop other sources of value-added through the private sector.[26]

The GCC economies are passing through a favourable stage that may not last for long. It allows the economy to benefit from the outcomes of large investments carried out by the government in fields of infrastructure including roads, ports, and logistic facilities. The governments in the region have also concluded several international and regional agreements, such as agreements on prevention of double taxation and others. GCC states are also members in a number of multi-lateral agreements, international and regional organizations have a number of bilateral agreements and have an A+ rating from some recognized rating agencies.[27] All these need to be best utilized for the benefit of the region, in particular regarding the private sector development and foreign investments. For that to happen, the relations among GCC economies need to be complementary instead of competitive. Having said that, there is a need for an institutional focus on competiveness: right now this is spread among several institutions. These countries must be able to develop specialization in products and services with comparative advantage through concentrating on the backbone or enabling sectors in terms of infrastructure, telecommunications, administrative services, and capacity development. Moreover, there is a need to improve the current governance practice in relation to public investment and public enterprises, consolidation of public investment companies and funds to produce clearer criteria for sector-based investments. Much has been done to upgrade the GCC economies and improve people's lives in the region, but much still remains to be done by both governments and the private sector. Government has a critical role in creating employment, improving health and education, and maintaining the environment and leading the transition toward well-functioning markets, with competitive and innovative businesses, rising productivity and incomes.[28] The private sector also has a key role in supporting inclusive growth, creating jobs, and providing critical basic services and public goods. Having said that, it is the private sector's responsibility to be better organized, more inclusive, more creative and more dynamic, in order to be a credible partner of government and change expectations for the better. All the above must be dealt with in a sense of strategy and, prioritized them based on what is important at each stage and in local circumstances.

Notes

1 Inclusive growth refers to a focus on economic growth that is both broad-based across sectors and inclusive of the large part of the country's labor force.
2 If managed well, FDI could bring a broad range of benefits to GCC countries.
3 MNCs are considered one of the most ways by which developing economies effectively become integrated to the global economy.
4 The Private sector is a critical partner in economic development, a provider of income, jobs, goods, and services to enhance people's lives.
5 Private firms and entrepreneurs invest in new ideas and new production facilities.
6 Share of Foreign Trade in the GDP, reached 95% in most GCC states.
7 The German think tank transformation BTI index (the Bertelsmann Transformation index), which analyses political and economic transformation, shows that all GCC countries have a BTI index ranging from limited to very limited.
8 The objective of joining multilateral and bilateral agreements is to support the private sector through encouraging foreign capital, which is usually accompanied with skilled labor, technology and access to the foreign markets. New agreements have been formed to facilitate access to new technologies, build partnerships with other countries in entering new markets and lower costs of production. However, in order to fully benefit from these agreements, the GCC private sector needs to be better prepared.
9 On the positive side, countries in the region actually started exporting more products over the last decade. Liberalization and greater openness increased the capacity of the domestic private sector to export, with a more diversified basket of goods.
10 Creating jobs for a young and better-educated labour force is a top priority of all governments.
11 The growth in this region is supported by oil revenues and fiscal expansion (government spending). Statistics indicate that the region need oil price to be at S 90 per barrel to meet spending commitments, which is alarming.
12 Thanks to higher oil prices.
13 They have concluded several international and regional agreements, such as agreements on prevention of double taxation and others. Also they are members of several multi-lateral agreements, International and Regional Organizations and havie an A+ rating from all recognized rating agencies.
14 Although it was important for the local manufacturing sector and is attractive for foreign investment. The cheap supply comes at a long-term cost. Some countries in the region will be facing energy deficits and will be net importers of energy in the near future and considering the use of more sustainable sources of energy such as renewables.
15 Educational attainment is raising the aspirations of nationals, but not necessarily their marketable qualifications
16 Government salaries are far higher than in the private sector. Although local people are generally paid better in the private sector than expatriates.
17 It is estimated that 45% of Saudi Arabia's government spending goes to cover public sector salaries.
18 Government salaries are far higher than in the private sector. Although local people are generally paid better in the private sector than expatriates, the majority remain lower paid than in the government.
19 Such as measured by GDP.
20 In a small open economy having the peg works as the strongest source of stability, which is so essential for promoting trade and investment in these countries.
21 SME-specific segments of banking Banks have limited incentive to engage on the SME side when other options are more profitable and less risky
22 Linked with increase in Oil price
23 For example; Private saving rates hovered at 4.8 percent on the average from 2002 to 2008.
24 Critical to understand the interactions among these elements.
25 In state business relation, Loyalty and alignment of interests is critical, as soon as personal agendas are involved, efficiency will be affected.

26 Find way to disconnect the clear relation between government spending and rest of the economy.
27 Many countries have opted to create an attractive investment environment, and improve their ranking in reports published by credit rating agencies and major international institutions such as Doing Business report issued by the IFC and the World Bank, Human Development Report (Rate during the past four decades) issued by the United Nations, World Competitiveness Report issued by the International Institute for Management Development (IMD), Switzerland, The report of international competitiveness of the World Economic Forum, Credit Rating Report issued by Moody's, World Economic Freedom Report issued by the Fraser Institute in Canada, Economic Freedom Report issued by the Foundation Heritage Foundation and Wall Street Journal, Transparency International report issued by Transparency International and others.
28 In terms of high cost of telecom and lack of integrated transport strategy. Also, electricity demand-supply poses challenge, as well as insufficient attention to business and professional services.

References

Al-Barwani, Khalfan (2008), Oman Country profile, The Road Ahead for Oman, Economic Research Forum.

Almezaini, Khalid S. (2012), the UAE and Foreign Policy: Foreign Aid, Identities and Interests, culture and civilization in the Middle East, Rout ledge

Baabood, Abdullah (2009), The Growing Economic Presence of Gulf Countries in the Mediterranean Region, Mediterranean Politics, Middle East

Booz & Company (2008), Youth in GCC Countries Meeting the Challenge

Chatham House (2012), Political and Economic Scenarios for the GCC, Middle East and North Africa Programme: Future Trends in the GCC Workshop Summary

Cohen, Stephen D. (2006), Multinational Corporations and Foreign Direct Investment, Oxford University Press

Cohen, W. and Levinthal, D. (1990): Absorptive Capability: a New Per- spective on Learning and Innovation, Administrative Science Quarterly, Vol. 35, pp 128-152.

Collombier Virginie (2011), Selected Players for Selected Reforms, an integrated report, ARI Thematic Study: Private Sector and Reform, Arab Reform initiative

Commission on Growth and Development (2008), The Growth Report: Strategies for Sustained Growth and Inclusive Development, Washington, DC: World Bank

Fasano, U. and Iqbal, Zubair (2003), AGCC Countries: From Oil Dependence to Diversification, IMF, Washington

Fasano, Ugo and Qing Wang (2001), "Fiscal Expenditure Policy and Non-Oil Economic Growth: Evidence from the GCC Countries" IMF Working Paper

GCC, Secretariat General (2012), "GCC: Twenty Years of Achievements", Riyadh

Giacomo, L. and Hertog, S. (2010), Has Arab Business Ever Been, or Will it Be, a Player for Reform?, ARI Thematic Study: Private Sector and Reform, Arab Reform initiative, Policy Paper 1 of 2

Gorg, H. and Greenaway, D. (2004). Much ado about nothing? Do domestic firms really benefit from foreign direct investment? World Bank Research Observer, 19(2): 171-97

Hammad, Alam, and Washington, Charles W. (1986), "Budgeting and Planning as Tools for Successful Development in Oman", America – Arab Affairs

Herb. M, (2008), Parliaments Economic Diversification and Labor Markets in Kuwait and
The UAE, manuscript.

Hertog, Steffen (2010), Benchmarking SME Policies in the GCC: a survey of challenges and opportunities, A research report for the EU-GCC Chamber Forum project

Hertog, Steffen (2012), Diversified But Marginal: The GCC Private Sector as an Economic and
Political Force, the Center for Gulf Studies at the American University of Kuwait

Hertog, Steffen and Luciani Giacomo (2009), Energy and sustainability policies in the GCC, Kuwait Programme on Development, Governance and Globalisation in the Gulf States

Hertog, Steffen and Luciani Giacomo, Valeri Marc (2013), The Politics of Business in the Middle East After the Arab Spring, Hurst, 2013

Hvidt, Martin (2013), Economic Diversification in GCC Countries: Past Record and Future Trends, Research Paper, Kuwait Programme on Development, Governance and Globalisation in the Gulf States

Mishrif, Ashraf (2010), Investing in the Middle East, Tauris Academic Studies

Peterson, J. E. (2009), Life after Oil: Economic Alternatives for the Arab Gulf States, Mediterranean Quarterly,

Ramady, Mohamed A. (2012), The GCC Economies Stepping Up To Future Challenges, Springer

Sager, Abdulaziz (2007), The Private Sector in the Arab World – Road MapTowards Reform, Arab Reform initiative

Smith, Adam (1776), The Wealth of Nations, W. Strahan and T. Cadell

Stiglitz, Joseph E. (2002), Globalization and Its Discontents; Allen lane, the penguin press

UNCTAD (2011), World Investment Report 2011: Non-Equity Modes of International Production and Development, York and Geneva: UN

United Nations (2012), Economic Commission for Africa, Unleashing the Private Sector in Africa Summary Research Report

World Bank (2005), World Development Report 2005: A Better Investment Climate for Everyone, Washington, DC: World Bank

World Bank (2009), From Privilege to Competition Unlocking Private-Led Growth in the Middle East and North Africa, Mena Development Report.

6

Awareness-DRM-Axis: Suggested Approach for Sustainable Development in the Gulf

Nilly Kamal Elamir

1. Introduction

The chapter argues that without sustainability any development model is incomplete and without a clear Disaster Risk Management (DRM) system all development achievements are at risk and without responsible individual and well established community awareness and communication channels at all levels any given country will not be able to introduce an integrated developmental model. Not to mention that integrating DRM into development is increasingly adopted by international organizations as well as governments in order to protect people and economy from impacts of natural hazards.

Meanwhile, at the Gulf Region studies, there is a lack of literature that links the natural and environmental characteristics with development. From that view, this chapter tries to explain the types of natural and environmental threats in the region by looking at environment crises in the Gulf, then the chapter will examine how these threats are considered on the awareness level and to what extend the DRM is used in the Gulf to combat such challenges. Also, the chapter will deal with recommendations towards a more balanced and Eco-system in the region using the DRM tools for sustainable development in the Gulf.

In addition to the above argument, the fact is the region has witnessed a number of oil spills with serious environmental impacts on the marine life which is the case also in other regions such as British Petroleum in the Gulf of Mexico of the United States. Such examples should be the base for assessing the environment and to set DRM mechanism in the Gulf region with a vision of making gulf cities safe and resilient ones free of oil spills using the GIS and other related techniques.

Taking into account that studying environment in a certain country or region requires an interdisciplinary background by definition, since topographic, social and cultural factors usually come together to explain any given environmental problem or its management the scale of the chapter is broad in some places. Similarly, looking at DRM and its applications in the Gulf has led to utilize a number of specialized concepts from applied science in some other cases.

2. Key Characteristics of Environmental Threats Gulf Region

2.1. State of Environment in the Gulf Region

An apparent climatic divide was identified for the Arabian Peninsula viz. 'Hot-Dry' and 'Hot-Humid' zones (Al-Khaiat, 2007, p.2). in respect of the marine environment, coral reefs are the most diverse environment of the marine realm. Not to mention that these are not only important biodiversity batteries, but also important for fisheries which used to be one of the main traditional economic activities in the region. While the mortality of a part of the coral reef system may have somewhat decreased the number of fishes (Dawoud et al., 2012).

The Gulf region enjoys a unique coastal environment, in part influenced by a shallow sea with unusually warm waters, (World Conservation Monitoring Centre, N.A), however, water scarcity is one of the main characteristics of the Gulf region. This has led to intensive dependence on desalination as a main source for fresh water supply for the domestic sector, since the mid-fifties of the last century. Most of the desalination plants are combined with power plants for power production. The main producers in the Gulf region are the United Arab Emirates (35% of the worldwide seawater desalination capacity) (Dawoud et al., 2012, p.3).

Further, climate change has an impact on the frequency of weather events in the region. High variations of rainfall with an increase in flood events impact the Arab region. For example, floods affected Saudi Arabia and Yemen from 2008 to 2009 with estimated economic damage of 1.3 billion USD. Tropical cyclones are another hazard in the Gulf region. Until 2007, the region was not considered prone to cyclones, when the Cyclone Gonu struck the Arabian Peninsula, and especially Oman (UNISDR-ROAS, 2013).

Natural and manmade factors combine to shape the environment system of the Gulf region. Desalination processes in Gulf Cooperation Council (GCC) countries similar to any other industry have their negative impacts on the Gulf environment. Some of these effects are impingement and entrainment of marine organisms due to the intake of seawater, the Green House Gases (GHG) emission due to a considerable energy demand of fossil fuel, and brine water discharge to the marine environment. Copper levels were reported caused by industrial outfalls or oil pollution (Dawoud et al., 2012).

It is worth mentioning that desalination technology was introduced to the region for the first time in the mid-fifties and has developed very rapidly to counteract the shortage and quality deterioration in groundwater resources and to meet more than 56% of drinking water supply (Al-Zubari, 2009).

Other environmental effects such as transport of suspended and bottom sediment and seawater stratification has been reported. Also, the seawater intakes are at the risk of pollution caused by oil spills, chemical pollutants and city drains and sewers that introduce a potential risk of ingress of such waters to desalination plants (Abdul Aziz, et al., 2000).

At the policy level, some actions have been taken aiming at protecting the Gulf environment. Many of Gulf countries have enacted national legislation for marine conservation and have signed various regional and international environmental agreements. However, the limited awareness of the importance of ecosystem services drawn from the coastal areas and their significance in sustaining economic growth has led to limitation of implemented policies and actions that engender better coastal management. (World Conservation Monitoring Centre, 1992)

2.2. Main Natural Crisis and Economic Related Environmental Issues in the Gulf Region

Moving to the natural threats at the Gulf countries level, the region is in fact vulnerable to floods, earthquakes, storms, and other hazards that killed approximately 3.3 million people between 1970 and 2010, average of 82,500 deaths worldwide in a typical year (The World Bank, 2010).

On the other hand, oil spillage is considered as the main manmade environmental problem the Gulf region faces.

However, the literature review for environment status in the Gulf has shown lack of studies on environmental system in many of the Gulf countries, as the main share of studies found was on Saudi Arabia compared with other Gulf States especially Oman on which environmental studies are unavailable in many cases.

In that context, the earliest recorded crisis in Saudi Arabia was the 1964 rains. Heavy rains poured continuously on parts of the country leading to a flood that killed 20 people and left about 1,000 people either injured or homeless. Other recorded crises are fire incident in Hajj season 1975 and Ras al-Khafji thunderstorm in October 1982 with over 200 deaths. More recent crises will be also mentioned in this section. (Alamri, 2010).

Compounding this problem is the geography of some of the most populated cities in Saudi Arabia such as Jiddah and Makkah, are on low ground and are surrounded by mountains. At the time of rain falling on these mountains, water runs in valleys towards these cities. With poor drainage systems, this continuous flow of water could easily lead to a flash flood. In 2009, Saudi Arabia severely suffered from flood which killed 163 people. Other major floods happened in 2005, 2003, 2002 and 1985 with fewer victims estimated at 126 persons in total. (Alamri, 2010)

Rising sea levels is considered another threat for the Gulf environment. Qatar, and the UAE are exposed, as over 2% of their respective GDPs are at risk from a Sea Level Rise of 1 metre, rising to between 3 and 5% for Sea Level Rise of 3 metres. Saudi Arabia and Yemen have more than 80 threatened animal species. Climate change could alter the animal composition of entire ecosystems. (Arab Forum for Environment and Development, 2009).

The lack of quality, quantifiable environmental data has proven a major hindrance to the global process of achieving sustainable development. The result is that throughout the world socio-economic and environmental decision makers are being challenged to make vital decisions without the necessary data and information (Arab Forum for Environment and Development, 2009). The Gulf region is no exception in that matter.

Unfortunately, there is no official publicly-available database that keeps a record of disasters in the Gulf. Surprisingly, most official information available comes from newspapers local to the region where the disaster occurred. Alternatively, databases usually set by international bodies such as International Disaster Database (IDD) of the WHO provides the best record of disasters in the region (Alamri, 2010).

"Historically, about half the oil transported through the global marine environment has come through the Gulf and the annual input of oil to the gulf's marine environment is skewed toward sources connected with marine transport. Currently, oil supplies about 40% of the world's energy, and recent estimates suggest that the Middle East has over 65% of world's petroleum reserves. This made oil spillage in the Gulf has been apparently heralded as one of the worst ecological catastrophes in history"(Alamri, 2010, p.8).

Seki accident, happened on 30 March 1994, is the largest oil spill recorded. (World Conservation Monitoring Centre, 1992) Because of its consequences and the uniqueness of geographical and biological features of the Gulf region, it is likely to become the worst ever man-induced oil-related marine disaster. Damages from oil spill after collision between the crude oil tankers Baynuna and the Seki released approximately 16,000 metric tons of light Iranian crude oil into the coastal waters of the Emirate of Fujairah, United Arab Emirates (Al-Ghais, et al., 1998).

Besides the dramatic and well-publicized catastrophic oil spills, a substantial amount of oil enters the marine and coastal waters of the Arabian Gulf and the Gulf of Oman Natural seeps also constitute a significant contribution (Al-Ghais, et al., 1998). Apart from the oil spills, disturbing ecosystems may be significant sources of greenhouse gas (GHG) emissions (Al-Zubari, 2009).

Some social and demographic factors usually hinder the recovery of the disaster such as the high rate of illiteracy and language barriers among vulnerable populations. Education level is in fact part of the problem. Illiteracy leads to less effective risk-communication and under-appreciation of the power of disasters. This keeps people unaware and unable to read safety brochures or use the internet and other media resources for public announcements can have adverse consequences and place the population at a higher risk of being a victim of disasters (Alamri, 2010).

For example, during the rainfall that resulted in the flood in Jiddah in 2009, many people ignored warnings about using motor vehicles for unnecessary trips simply because illiteracy meant less attention to such messages. Some people under-estimated the risk and decided to take a trip in their cars to "enjoy" the rain, and these were the cars that were swept away by the flood and clogged main streets. (Alamri, 2010).

Similarly, it is important to reflect back on previous disasters and learn lessons from them to avoid committing the same mistakes again.

Aside from the negative impacts on the Gulf region environment, blue carbon comes as an opportunity for protecting the natural environment of the region. "Gulf region and particularly coastal areas of states in the Gulf are closely tied to Blue Carbon ecosystems through cultural heritage, their role as nursery grounds for fisheries, and through coastal development. There are risks and uncertainties that need to be considered when advancing Blue Carbon in the region (the stored carbon in coastal ecosystems can be lost through the impact of climate change itself and changes in land use both along the coast and further inland). There is still uncertainty about the amounts sequestered in each Blue Carbon ecosystem – research is required to generate these data and reduce the uncertainty" (Environment Agency Abu Dhabi, 2011, p. 8).

3. Environmental Awareness in the Gulf Region: Does it Reflect the Threats?

As mentioned in the introduction, understanding the environment in one region requires linking natural characteristics with the people's understanding of it, or understanding the level of people's awareness because they are the base for any development process.

This chapter argues that strong awareness in a given society is an outcome of the effective role of three parties: the individual, the scientists and the policy makers.

3.1. Awareness in the Gulf Region: Science and Technology Level

Reviewing the literature and policy towards environment in the Gulf shows to a great extent early attention to the issue even it was not on a large scale. Since the 1990s, some Arabic books attempted to establish the linkages between environmental and developmental issues. In 1990, the United Arab Emirates University published the proceedings of a symposium on the economic and environmental dimensions of development in the GCC countries (United Arab Emirates, 1990). This was followed by a book by Al-Kandary (1992) published by Kuwait University entitled Environment and Development and other writings alerted the Arabs to the developmental costs of neglecting environmental issues (Selim, 2004).

However, the question remains whether these books have comprehensively covered the state of environment in the Gulf as well whether these efforts have successfully reached policy makers with solid and practical recommendations for protecting the environment.

In other words, the role of scientists is indispensable as environment assessment makers and solution providers. Thus, research work on the Gulf needs to be originated in the region rather than imported from the science production in the West.

The Gulf region needs studies that shed light on concepts like resilience as well as DRM and making linkages with social science and the role of citizens and state. Resilience is meant here as "a community or region's capability to prepare for, respond to, and recover from significant multi-hazard threats with minimum damage to public safety and health, the economy, and national security" (Cutter, 2008, p.2)

Another important concept, as mentioned in the previous section, is blue carbon which refers to the coastal ecosystems including mangrove forests, seagrass meadows and saltwater marshland, are gaining increased attention for the carbon stored in their biomass and sediments (Environment Agency Abu Dhabi, 2011).

Similarly, an ongoing coordination has been started a few years ago within the Arab countries for example, 'Impact of Climate Change on the Arab Countries' is the second in a series of annual reports produced by the Arab Forum for Environment and Development (AFED). The first AFED report, published in 2008 under the title 'Arab Environment: Future Challenges', covered the most pressing environmental issues facing the region, and went beyond to provide a policy-oriented analysis. The report was presented to AFED's annual conference which convened in Manama in October 2008 (Arab Forum for Environment and Development, 2009).

3.2. Awareness in the Gulf Region: Individual and Civil Society Levels

If the community level is missing the state level will lack communication with people and will be only an up bottom policy.

This individual level in fact compromises the citizens' behaviors, private sector's initiatives and civil society contributions towards raising the people's awareness. At the Gulf level, some private sector firms adopted environmental awareness programs as part of its corporate social responsibility. This is done through adopting strategy that promotes environmental awareness among employees and their families, general public, and local industries. The environmental awareness activities associated with this strategy are designed to enhance the environmental knowledge of employees the community and local industry to cultivate an environmentally responsible culture. Similarly, some other firms announced their commitment to powering economic development with adequate and affordable energy as well as a low carbon future (Saudi Aramco, 2014).

In a survey conducted in 2008 in 128 countries by the Gallup Poll, 49% of the surveyed individuals in Saudi Arabia were aware of climate change and 40% perceived it as a threat to their country, while 39% believed that it was caused by human activities (Darfaoui, et al., 2011).

Thus, only under half of the people in the biggest Gulf country recognize the environmental threat, where the Gulf as a region is considered as the poorest region with respect to water resources, with almost no rainfall and no groundwater. With the exception of Yemen, the rest of Gulf countries heavily depend on the desalination of water from the Gulf. This water scarcity should be reflected in people's level of awareness. It should be also linked with climate change problems including drought severity as serious challenge.

Nonetheless, the rate of solid waste generation in Bahrain is almost 2.7 kg per capita/ day as a result of growing family income, increased purchasing power alongside flourishing urban and trade businesses (Darfaoui, et al., 2011) where it is only 1.1 kilo in Japan (Ministry of Environment, 2012).

The civil society role is limited in the Gulf but it should take the lead towards designing programs to teach dedicated skills necessary to solve present and future environmental problems. Prepare community educators for the future able to use scientific methodology for collecting, summarizing and analyzing data, especially that related to local, regional and global environmental problems. Enhance interdisciplinary programs with emphasis on achieving a sustainable-earth society, conservation of resources, monitoring environmental pollution, waste management, and risk analysis as well as its assessment (United Arab Emirates University, 2010).

Emphasizing community education in disaster management in the Gulf: to aim at increasing public awareness about potential hazards and threats and mitigation measures people can undertake about reducing their impact. Such a program will face serious challenge which is required community and strong nongovernmental organizations and active civil society to be organized together with hospitals, fire brigade and police good strategy for communication is to provide information in a simple, easy-to understand language upon the most feasible mitigation solution (Vasta, K., 2003).

3.3. Awareness in the Gulf Region: Policy Action Level

Any solid action should include specific objectives and implementation plans within a fixed timeframe. At Arab and Gulf levels, a number of actions have been agreed upon to protect the environment even if the awareness component was not clearly mentioned.

One of the first actions in that field was the workshop held on biological diversity in the GCC jointly with the World Conservation Union /GCC in Kuwait in September 1994 together with Kuwait's Environment Protection Council. The workshop reviewed implications of the Convention on Biological Diversity (CBD) for GCC countries and specify requirements for implementation (Aspinall, 2001). Unfortunately, progress in these directions remains unclear.

Another example at the policy level is concerned with waste management. The GCC member countries devised a uniform waste management system that was adopted in December 1997 to codify waste treatment including domestic, commercial or industrial, inactive or hazardous. This action targeted put in place a monitoring mechanism for waste production, storage, transport, treatment and disposal. Prosperity of humans and ensuring long and short-term environment protection were also among these coordination objectives (Darfaoui, et al., 2011).

Some other specific examples are the first two Arab green building councils in The UAE and Egypt; the massive forestation program in the UAE; Masdar, the first zero-carbon city in Abu Dhabi. King Abdullah University of Science and Technology (KAUST) in Saudi Arabia has been established as a centre of excellence on energy studies; such initiatives are perfect tools of transforming oil income into future technology. There is also AFED's Arab Green Economy Initiative, an exercise in public-private partnership (Arab Forum for Environment and Development, 2009). Still, it is essential that these initiatives become part of an integrated, large and sustainable development plan, in order to move from fragmented initiatives to a comprehensive policy framework at the national level, let alone at the regional one.

Arab-Arab cooperation can also be improved as a means for better environmental policy in the GCC. Some steps have been already accomplished, for example in the areas of energy efficiency and renewable energy, the use of compressed natural gas as a transport fuel, and investing in carbon capture and storage. These steps have been taken at the individual country level and need to be integrated into regional level for maximum benefit.

The Council of Arab Ministers Responsible for the Environment (CAMRE) issued a landmark Declaration in 2007, which adopted the scientific consensus that was reached by the IPCC, accepting that the increase of global temperatures was mainly due to human activities (Arab Forum for Environment and Development, 2009).

At the international level, role of the United Nations Educational Scientific and Cultural Organization (UNESCO) jointly with Arab League Educational, Scientific and Cultural Organization (ALESCO) is to be mentioned. The two parties include environmental matters of mutual concern and transformed this concern into tangible outcome. The Regional Organization for the Protection of the Marine Environment (ROPME), with a secretariat in Kuwait, is part of UNEP's Regional Seas Programme. The World Conservation Union (IUCN) is also active in the region, independently and in collaborative efforts (Aspinall, 2001).

Also, AFED conducted a pan-Arab survey aimed at exploring awareness of climate change among the Arab public, and their willingness to personally contribute to climate change mitigation. The results showed increasing awareness: 98% believed that the climate is changing, and 89% believed this was due to human activities. 51% believed that governments were not acting adequately to address the problem, while 84% believed climate change posed a serious challenge to their countries (Arab Forum for Environment and Development, 2009).

Conducting polls and surveys are essential tools for awareness assessment. Thus, Gulf States need to conduct periodical surveys for measuring awareness levels and develop policies and awareness programs based on these surveys results.

Similarly, making people aware of healthy ecosystem requires clear identification of gaps in legislation and the will to create and enforce a strong legal and institutional framework, including economic incentives to reinforce desired behaviors and outcomes.

Also, it is important to mention that, internationally well-established guides have been set. UNISDR has launched the "Making Cities Resilient" campaign to support urban areas to become more resilient to disasters. Almost 300 cities and municipalities in the Arab region have joined the campaign (20% of all cities worldwide). The Mayors Handbook on "How to Make Cities More Resilient" was translated to Arabic and has been disseminated widely in the region (UNISDR-ROAS, 2013).

Media production either public or private sector ones can play a role on introducing and marketing the healthy natural coastal ecosystems, such as mangrove forests, saltwater marshlands and seagrass among people. These productions should also focus on benefits including ecosystem services such as a rich cultural heritage; the protection of shorelines from storms; erosion or sea-level rise; food from fisheries; maintenance of water quality; and landscape beauty for recreation and ecotourism. In a "Blue Carbon" context these ecosystems also store and sequester potentially vast amounts of carbon in sediments and biomass (Environment Agency Abu Dhabi, 2011).

4. Types of DRM Mechanisms Required for Better Development and Environment in the Gulf

4.1. Achieved Progress on DRM in the Gulf

No community can ever be completely safe from natural and man-made hazards, it is useful to think of a disaster resilient community as 'the safest possible community that we could design and build in a natural hazard context minimizing its vulnerability by maximizing the implementation of disaster risk reduction measures. In other words, DRM is required and needs to be the baseline of policy drawing in the Gulf not only for the flood management but also for each single economic activity as the environmental condition alerting in many fields.

While numbers and information are starting points for solid DRM, databases are incomplete at the regional level. We can admit that there is lack of official publicly-available database that keep a record of disasters in the country. Most official information available comes from newspapers local to the region where the disaster occurred (Alamri, 2010).

Despite this, some Arab countries have started to report on their disaster losses hopefully to provide a basis for informed risk analysis and the development of disaster risk reduction policies. Nine out of the 22 Arab countries have either completed or initiated the development of national disaster loss databases. (UNISDR-ROAS, 2013) Unfortunately, none of these countries are from Gulf region.

However, the last few years witnessed willingness to reduce impacts of disasters and to be better prepared. Some achievements have been attained in the Arab region. At the political commitment level, Heads of Arab States adopted the Arab Strategy for Disaster Risk Reduction 2020, under the auspices of the League of Arab States. The Strategy had been adopted by the Council of Arab Ministers Responsible for the Environment (CAMRE) and the Socio-Economic Council of the League of Arab States in 2011 (UNISDR-ROAS, 2013).

For the first step of its kind, the GCC, in 2013, committed itself to take steps to develop a risk reduction road map. The Secretary-General of the Cooperation Council for the Arab States of the Gulf has called for strong regional commitment towards development of a disaster risk reduction strategy to strengthen the resilience of nations and individuals to natural hazards (UNISDR-ROAS, 2013).

4.2. Requirements for Better DRM in the Gulf Region

In this section, the chapter introduces a number of specific areas where the DRM policy should be drawn to make the Gulf States safer and more resilient. Some of which have been introduced by the World Bank literature on DRM as universal guidelines for countries to adopt strong DRM programs.

Since the best DRM requires sustainability as well as development, and since the construction work consists one of the main activities for the GCC economies, demanding certain provisions for a design code emerges as a critical need to cater to local challenges in concreting (Al-Khaiat, 2007).

The concept of disaster risk management (DRM) is new to the Gulf States. As a result of a current focus on post-disaster relief, they have relied largely on central government agencies to mobilize for relief activities.

With this view in mind, "an apparent climatic divide was identified for the Gulf viz. 'Hot-Dry' and 'Hot-Humid' zones which were further classified into an order of exposures, detailing the potential dangers to concrete durability. It is intended that this contribution would help formulate a design code for concrete durability in the Gulf. In other words, the region needs to develop guidelines on building and fire codes as part of risk mitigation. This also includes basic and infrastructure and structural measures, such as, construction of dams, levees, flood control structures etc." (Al-Khaiat, 2007, p. 5).

Another vital part here is sustainable food self-sufficiency in the region. The GCC countries depend heavily on importing their food and minor portion of their needs are met by domestic production, which does not exceed 20% or even less. This leaves countries exposed to price risk either relating to volatility of import prices or and supply risk. Not to mention the tensions in the Arab and Middle East regions that may affect the trade flows via Hormoz strait, thus GCC governments are in a serious need to hedge supply risks through strategic storage and investments in port and transportation to create a regional import and transport network.

Developing such strategies should also come as part of national or regional vision for locally-driven capacity development strategy, through which an umbrella organization are established to conduct research and advise policy makers on (gradual) self sufficency for food production in the long term. The region needs to prepare national calibers capable of assessing all environmental impacts using comprehensive risk assessment matrices. Advanced 3-D modeling techniques can be used to simulate the water hydrodynamics and oil weathering and transport in the Arabian Gulf and there is a need to be used in a wider scale. (Elshorbagy, 2008).

Similarly, the GCC as a regional organization needs to adopt a vision to coordinate Participatory Risk Assessment towards its first step which is hazard analysis. Hazards Analysis shows the potential impact of a hazard on a geographical area. It involves seeking relevant information upon all the hazards to which a community is exposed (Vasta, 2003).

Upon the assessment completion, and a disaster risk profile for each of the gulf countries, GCC should work, preferably collectively, on designing effective National Disaster Management Systems. National Disaster Risk Management Systems include changing the attitude of governments from reacting with policy after disasters to take a more proactive form of emergency management. This may be drawn from country experiences and recommendations from related research work. Successful national systems should have provisions to ensure sufficient resources for key players to carry out their responsibilities.

This national system should ensure that disaster risk reduction is a national and a local priority with a strong institutional basis for implementation; identify, assess and monitor disaster risks and enhance early warning use knowledge, innovation and education to build a culture of safety at all levels. (El Raey, 2011)

Another essential component for the DRM system is early warning systems of extreme events such as flash floods and dust storms must be established. Preparedness and mitigation strategies, combined with high coping capacity ensure that, although events may cause extensive damage, they have been planned for.

Here, the South American countries can be a reference. El Salvador's National Civil Protection System includes National Commission's duties for designing the National Policy for Civil Protection and Disaster Prevention and Mitigation; proposing to the declaration of a State of Emergency and providing immediate response and keeping public order; supervising the implementation of Disaster Prevention and Mitigation Plans.

However, most of proposed policies might seem ambitious and challenging, so, cooperation with UN agencies working on DRM can be an asset especially at the first stages, and one of the main UN active agencies in that field is the United Nations Office for Disaster Risk Reduction (UNISDR).

UNDSIR stated the challenges for the future remain to be addressed which typically applied to the Gulf country level include national level coordination: The ownership of disaster risk reduction requires clarification the national level. It is important to better define roles and responsibilities amongst national bodies to ensure that DRR is addressed comprehensively and effectively (UNISDR-ROAS, 2013).

In the same context, including the internationally agreed works in the policy making process in the Gulf may be another asset. Atop of these works is the Hyogo Framework which explains mechanisms for international and regional cooperation and assistance in the field of disaster risk reduction through, transfer of knowledge, technology and expertise to enhance capacity building for disaster risk reduction; sharing of research findings, lessons learned and best practices; compilation of information on disaster risk and impact for all scales of disasters in a way that can inform sustainable development and disaster risk reduction; for awareness-raising initiatives and for capacity development measures at all levels (UNISDR, 2005).

Financing DRM systems or reconstruction after any disaster is costly as well as complex. The Gulf States might find it useful to exchange experiences and knowledge with specialized international institutions in disaster risk reductions and management such as GFDRR together with United Nations, European Union (The World Bank, 2010). These institutions amongst others provide technical assistance and fast-track financing for recovery and reconstruction planning with the fund coming from the GCC side.

On the Gulf side, it is important also to set measures for protecting mangroves as a tool for mitigating vulnerability to Sea Level Rise. Despite the fact that they are found scattered along much of the Red Sea coast, the major concentration is in the southern red sea where factors such as increased sediments create an environment more conducive to their development. Agricultural development, properly planned and managed, could be beneficial to certain coastal habitats such as mangroves. Mangroves have a variety of values: they provide food in the form of detritus and shelter for numerous organisms (such as shrimps, and fish) (El Raey, 2011).

Communities in the Gulf seem to be experienced in crisis management since long time ago. The management of potentially hazardous situations such as religious mass gatherings has been the duty of the people of Makkah for many centuries. Inhabitants of Makkah used to evacuate their houses to accommodate the incoming pilgrims, and servants of the Holy Mosque used to distribute cold water to quench pilgrims' thirst. This concept of serving mass gatherings formed the nucleus of the first emergency management plans in Saudi Arabia (Alamri, 2010).

However, implementation of effective disaster management systems requires political commitment, which manifests in appropriate policies, planning, supporting legislation and resources devoted to disaster management issues. While the systems vary according to political and economic culture of the countries, there are some general requirements to be considered when designing the governance systems. We need policy makers as well as public understand that the occurrence of extreme catastrophic events can be sudden, quick and unpredictable The World Bank, 2003). So, prevention is possible if we rethink the way we live and use our resources more efficiently at the time of welfare so we will be relatively prepared at the time of risk.

5. Recommendations for Sustainable Development in the Gulf Region

Here, regional cooperation is to be mentioned, for example, Yemen who suffers from environmental risks, so, through regional coordination on combating pollution the two sides can gain benefits mutually. Gulf countries can also develop cooperation policy in the field of environment and DRM with Asian countries who are also at environmental risk to strengthen ties with Asian partners and to verify the ally of the region. Not to mention the progress that many Asian countries achieved in the renewable energy Research and Development and the Gulf can learn from and can transfer that knowledge into giant economic projects.

Reaching that level of safety needs an advanced level of awareness and part of that is to make huge investments in the remote sensing incentives for environment and science parks and green hotels.

Not to mention that number of challenges are there if we want to raise the awareness the multicultural environment in the gulf countries on issues of comprehensive risk assessment for the gulf region environment: water resources, sea level rise, human health, food security, tourism, land use and urban planning and biodiversity.

At the international level also, Gulf countries should investigate cooperation with other partners such as UN agencies especially to reach know-how and expertise on solar, wind, geothermal, biomass energies which will minimize impacts on Gulf climate. This may be in the form of renewable energy driven desalination technologies or as compensation measures such as the installation and use of renewable energy in other localities or for other activities.

However, a plan is required to reduce the dependence on "imported" technical expertise, which often leads to sub-optimal outcomes. The United Nations system has been playing a central role in regional efforts to build capacity (World Conservation Monitoring Centre, 1992).

Alternatively, experience exchange between the Gulf countries is very important due to the climate conditions similarity. The importance of mangrove ecosystems in the UAE is already strongly acknowledged, with the total area of mangroves increasing through many afforestation efforts. Sea grass meadows are recognized for their importance as habitat for threatened sea turtles. These models can be transferred to other Gulf countries such as Saudi Arabia who has intensive mangrove ecosystems.

Another important policy level for better DRM and environment in the Gulf is legislation improvement and update. The Gulf countries should work together to adopt strong legislations on disaster law to put a basis for disaster risk management and the role of the state in case of crisis. This is to include also distinguishing between risk management, emergency management and disaster management. Such laws should also establish institutional frameworks for people's attitude towards their environment such as waste management and open the door for more private sector investment in new and renewable energy, in addition to the system for dealing with risks and disaster either natural or manmade ones, such laws should set sustainability as the priority for any economic activity.

Science and technology serving sustainability is another important component to improve the DRM culture in the Gulf. Environmental science in the Gulf is a relatively new specialization. The policy trends in the region seem to be focusing on an economic model of cost-benefit analysis rather than sustainability. Such vision needs to be changed and to make the region adopt an environmental oriented policy.

Doing this, the governments in the Gulf should establish research and consultancy firms that reporting to the government and these units are to take care of all environmental issues. Such units are expected to undertake literature reviews of disaster risk management research work to draw lessons for the region. There is much research investigating using different renewable energy sources in GCC countries.

"Research indicates that Blue Carbon ecosystems have the potential to hold vast stores of carbon and are important for nature-based approaches to climate change mitigation. These research needs to be deepening and shared among the Gulf countries." (Environment Agency Abu Dhabi, 2011).

On the other hand, Gulf governments should encourage opening research units and university colleges on the environment as a major either for social or applied sciences to integrate with each other and help in bringing up next generation with solid awareness of the environment.

One of the missions for environmentalists in the Gulf should be how to link DRM literature with social science for a better prepared society for crisis, disaster risk reduction, and how to raise people's levels of awareness.

The region needs comprehensive security studies. These studies should look at environmental study as an essential part of national security and as important as the military ones. These studies should be balanced and covering all Gulf countries not only some of them.

6. Conclusion

This chapter has explained how the Gulf region is threatened by a number of environmental risks in addition to the rapid economic growth these two sides call urgently for applying the DRM and integrating it into to the development policy in the Gulf States.

Because protecting environment system is a responsibility of integrating circles including people, government and the community as a whole, some recommendations introduced and were divided into the three circles of individual, policy action and community.

From previous sections of the chapter, it was suggested that the Gulf States are prone to natural as well as manmade risks. It was clear also that dealing with these risks through solid DRM strategies has been started but yet to be solidly formalized.

Gulf societies need to adopt positive features of cultural globalization of less consumerism and more productivity, awareness as well as public participation. Environmental education should take place in the region with an emphasis on Islamic values as taught by Prophet Mohammed who said: "He who is secure in his house, healthy in his body and has his food for the day, has owned the world".

Regarding preparedness and awareness at time of crisis, Islamic values should be the reference especially in setting environmental curricula in the region. Some people have the attitude that "what God wills to happen, will happen"; however, this contradicts Islamic beliefs who ask people to take responsibility first. Islamic teachings state that every person has to do their best in taking precautions, as well as believing in God and relying on Him (Alamri, 2010).

References

Abdul Aziz, P. K., et al., (2000) Effects of Environment on Source Water for Desalination Plants on the Eastern Coast of Saudi Arabia", Desalination 132, 29-40.

Alamri, Y. A. (2010) Emergency Management in Saudi Arabia: Past, Present and Future: University of Christchurch, New Zealand.

Al-Ghais, S., Walter. H. P. (1998) Assessment of Damages to Commercial Fisheries and Marine Environment of Fujairah, United Arab Emirates, Resulting from the Seki Oil Spill of March 1994: A Case Study: Environmental Research and Wildlife Development Agency and Battelle Memorial Institute, Yale F&S Bulletin 103, 407-428.

Al-Khaiat, H. (2007) Designing Durable Concrete Structures in the Arabian Gulf: A Draft Code (Online) available from: http://cipremier.com/100032012 (Accessed: 12th February 2014).

Al-Zubari, W., (2009) Water Resources Management Issues and Challenges in the Gulf Cooperation Council Countries: Four Scenarios, in Cronin, R., Pandya, A., eds., Exploiting Natural Resources: Growth, Instability, and Conflict in the Middle East and Asia, Washington: Stimson.

Arab Forum for Environment and Development (AFED) (2009) Technical Publications and Environment & Development Magazine.

Aspinall, S., 2001. Environmental Development and Protection in the UAE. in: Al-Abed, I., Hellyer, P. (Eds). United Arab Emirates: A New Perspective. Trident Press, Bookcraft, UK. pp. 277-304.

Cutter, S., et al. (2008) Community And Regional Resilence: Perspectives From Hazards, Disasters, And Emergency Management, Report 1: University of South Carolina.

Darfaoui, A., and Al-Assiri, A. (2011), Response to Climate Change in the Kingdom of Saudi Arabia: FAO, unpublished paper.

Dawoud, A., Al Mulla, M. (2012) Environmental Impacts of Seawater Desalination: Arabian Gulf Case Study (Online) available from: www.sciencetarget.com (Accessed: 13th February 2014).

El Raey, M., (2011) Impact of Sea Level Rise on the Arab Region. Alexandria: Arab Academy of Science, Technology and Maritime Transport, unpublished paper.

ElshorbagY, W., and Elhakeem, A., (2008) Risk assessment maps of oil spill for major desalination plants in the United Arab Emirates, Desalination 228 , 200–216.

Environment Agency Abu dhabi (2011) Blue Carbon: First Level Exploration of natural Coastal Carbon in the Arabian Peninsula (With Special Focus on the UAE and Abu Dhabi): AGEDI.

Ministry of Environment (2012) Solid Waste Management and Recycling Technology of Japan Towards a Sustainable Society: Waste Management and Recycling Department.

Saudi Aramco CEO's Remarks at the UN Climate Summit 2014, available from: http://webtv.un.org/live/watch/industry-action-announcements-climate-summit-2014/3803270960001 (Accessed: 5th December 2014).

Selim, M., Environmental Security in the Arab World, Paper prepared for presentation at the Meeting of the International Studies Association, 17-20 March 2004, Montreal, Canada.

The World Bank, and GFDRR, (2003) Building Blocks of Comprehensive Disaster Risk Management: Concepts and Terminology: World Bank Institute.

The World Bank, and GFDRR, (2010) Natural Hazards, Unnatural Disasters – the Economics of effective prevention: The World Bank.
The World Bank, and GFDRR, (2010) Understanding Risks, Proceeds from 2010 UR Forum: The World Bank.
UNISDR, (2005) Hyogo Framework for Action 2005-2015: UNISDR.
UNISDR-ROAS (2013) UNISDR Factsheet: UNISDR.
United Arab Emirates University (2010) Faculty of Science Environmental Sciences Graduate Program: UAEU.
VASTA, K. (2003) Community based Disaster Risk Management, presentation to the E-Learning course on Disaster Risk Management: World Bank Institute.
World Consevation Monitoring Centre (1992) Gulf War Environmental Information Service: Impact on the Marine Environment. Cambridge: Cambridge University Press.

7

Advancing Sustainable Development in Bahrain through the Triangulation Approach

Latifa Al-Khalifa

1. Introduction

The Imperative Call for Sustainability. Government attempts for sustainability in all its jurisdictions have been given utmost significance in the recent years due to the manifest necessity to safeguard various forms of resources for the future stakeholders and to provide a route for development and growth that is rooted in the premise of positive conservation, protection, and continuity. These endeavors for sustainable development were put forward in a methodical process due to the creation of several frameworks for sustainability, deliberate interventions of international organizations, collaboration of various worldwide alliances, and bolstering support systems in various nations.

The significance of sustainable development is succinctly presented in its typical definition that is grounded on the concept to pursue a type of development that "meets the needs of the present without compromising the ability of future generations to meet their own needs" (World Commission on Environment and Development, para 1, 1987). Inherent in this conceptualization is the primacy of resource perpetuity to the next generation of stakeholders in a particular nation. The ramifications of social processes such as those related to commerce, trade, public governance, education, scientific innovation, and sociocultural processes are considered in the framework of sustainable development, which makes this advocacy an important pillar in all human activities.

Given the aforesaid importance of sustainability, it is easy to identify sustainable development efforts around the world. In North America, the "Federal Sustainable Development Strategy (FSDS) puts the Government of Canada's environmental priorities squarely within the broader context of social and economic priorities. The environment has an equal footing with the social and economic pillars of sustainable development" (Public Works and Government Services of Canada, p. v. 2010). Most of these efforts in sustainable development have been significantly accomplished and realized in larger economies. For example, "Germany, the United Kingdom and France have all managed to reduce their emissions over a 40-year period while their economies have continued to grow... Germany has reduced total emissions by 22 percent, France by 20 percent and the UK 18 percent" (Brown, sec 3, 2013). This is supported by the concerted efforts of the European Union (EU). The Eurostat European Commission reports that the "EU sustainable development strategy, launched by the European Council in Gothenburg in 2001 and renewed in June 2006, aims for the continuous improvement of quality of life for current and future generations" (Eurostat European Commission, p. 2, 2009). In the UK alone, "eight Regional Development Agencies (RDAs) were launched in 1999, with a ninth - for London – added in July 2000. The RDAs have a statutory purpose to contribute to the achievement of sustainable development in the UK" (Stratos Inc., p. 5, 2004). In relation to this, a holistic sustainable development strategy is reflected in the report of the International Institute for Sustainable Development, an organization which uses a publication as tool in "identifying key challenges faced in relation to the strategic management aspects of national sustainable development strategies including leadership, planning, implementation, monitoring and review, co-ordination, and participation of different countries"(International Institute for Sustainable Development, para 3, 2013).

Global Challenges in Meeting the Demands for Sustainability. Nonetheless, the implementation of sustainable development programs in different countries is confronted by many challenges that are primarily political, economic, and technological in nature. In the report of Brown (2013), it was explained that China, the US and India have failed to tackle the issues in sustainable development. These recorded failures may be attributed to certain factors that result to predetermined detrimental effects. Connor states that the "extraction, refinement, transportation and storage of fuels carries an immense environmental burden, as does its ultimate consumption, and disposal of waste products. These burdens have local, regional and global manifestations, ranging from impacts on soil, groundwater and land-use, to those on atmosphere and ocean"(Connors, p. 2, 1998).

The very nature of sustainable development sometimes poses innate contradiction to the supreme goal of commercialism and modernization. For development to occur in business, companies make use of various forms of energies which sometimes result in "climate change, energy consumption, waste production, threats to public health, poverty, social exclusion, management of natural resources, loss of biodiversity, and land use"(Legrand Group, sec 4, 2014). Such consequences are the exact opposite of the goals of sustainable development because "sustainable development implies the fulfillment of several conditions: preserving the overall balance, respect for the environment, and preventing the exhaustion of natural resources." (Legrand Group, sec 1, 2014).

The United Nations Department of Economic and Social Affairs (p. 2, 2013) enumerates challenges in the implementation of sustainable development, which includes "impact of climate change, [which] threatens to escalate in the absence of adequate safeguards; ... hunger and malnourishment; ... income inequality within and among many countries; ... rapid urbanization; ... [and] recurrence of financial crises." It is evident that despite the efforts of various organizations and governments around the world, advancement in the implementation of sustainable development continues to be hindered by many factors. McKeown (para 3, 2002) explains that "this lack of progress stems from many sources. In some cases, a lack of vision or awareness has impeded progress. In others, it is a lack of policy or funding."

Zooming into the Gulf Region's Sustainable Development Status. In the context of the Gulf region, these issues accompanying the local sustainability scheme continue to be bottlenecks of its effective and efficient promulgation. Sillitoe (2014) provides a substantial elucidation of this proposition by highlighting the Gulf's reliance on non-renewable forms of energies and processes to supply global energy needs. Elgendy (para 2, 2012) highlights the proposition that "embracing sustainable development in the built environment of the Middle East faces many challenges, which prevents it from becoming part of the region's development framework and its building industry practices." A common challenge in the region is maintaining sustainability of water resources (Smith, 2009). This perennial problem has continued to batter authorities' decisions especially in the aligning of government policies with the agenda of sustainable development.

Concerns related to sustainable development have also relevance in the context of Bahrain, one of the countries in the Gulf region, where dependence on non-renewable energies is evident and highly palpable. Aside from these concerns on energy sustainability, the issue on commercial sustainability also presses present-day policy makers. From this argument, it can be purported that other dimensions of sustainability are being confronted with challenges in the region. Environmental sustainability is not the only aspect of development that is faced with challenges and limitations, but also other interrelated areas of human and social activities.

In the Kingdom of Bahrain, programs on sustainable development are prominent in the nation's political and social processes. Its government has laid "sustainable development plans amongst its top priorities and set up a bundle of integrated programs aimed at raising the standard of citizen's living in terms of education, healthcare and social welfare as an embodiment of its vision"(Brittlebank, para 2, 2012). Such sustainable development agenda of the country reflects a comprehensive and coordinated attempt to include all aspects of social processes. Integration of all these processes in the overall sustainability plan is not only intrinsic to the accomplishment of certain goals but also an imperative function of the state. The account of the United Nations Development Program (UNDP) (para 1, 2012) in Bahrain supports the aforesaid integration process: "The Kingdom of Bahrain, in partnership with UNDP, is committed to empowering citizens to expand their use of available resources in order to meet their own needs, and change their own lives by integrating social inclusiveness, economic growth, and environment." This is operationalized through UNDP's coordination with other sectors of Bahrain such as the Ministry of Interior, Economic Development Board, Chamber of Commerce and Industry, the Ministry of Industry and Commerce, and other public and private organizations. Because of these coordinated efforts, a recent recognition was received by the government. Rafique (para 1, 2014) specifically reports that "His Royal Highness Crown Prince and first Deputy Premier to reinforce the sustainable development of the Kingdom, under the development program initiated by His Majesty King Hamad, as well as the government's efforts [to] deliver significant improvement to public services."

A Rationale for a Framework in Sustainable Development Implementation in Bahrain. Given these critical matters linked with the operation of sustainable development programs, this chapter provides a framework for reinforcing the efforts for sustainability in the Kingdom of Bahrain with the use of a threefold Triangulation Approach and with the primary emphasis of the roles of youth, women, and other sectoral dimensions of society. It has been argued in the aforementioned section that Bahrain, like other countries in the GCC, is confronted with various challenges in the implementation of its sustainable development agenda despite its deliberate efforts to coordinate and collaborate with local and international governing bodies. What remains to be the existential question is the citizens' awareness on these existing programs and strategies of the government to advance sustainable development. If such awareness exists, the next ontological query is the extent of such awareness among the stakeholders and how the government can enhance this collective consciousness to efficaciously implement the goals of sustainable development in the region. Determining the stakeholders' awareness on the sustainable development agenda of the country is not only practical but also imperative since program implementation can only be effective when the individuals and groups concerned are in regular and deliberate interaction with each other with the end of accomplishing a common interest. With this proposition, it is also a focal objective of this inquiry to formulate an approach in raising the sustainable development awareness of the stakeholders of Bahrain community.

Figure 1 depicts the discussion paradigm of this inquiry. As seen from the schema, the first section of the chapter documents the extent of sustainable development awareness of the stakeholders in Bahrain in terms of the youth, women, private sector, government sector, and the expatriate sector. At the onset it is hypothesized that there is a lack of awareness on the basic tenets of sustainable development among the stakeholders in Bahrain. A term called Sustainable Development Awareness (SDA), a novel terminology, is introduced anddescribed exhaustively in the literature review section of the chapter using Bahrain as the local environment. Formulation of this construct is based on the conceptualizations available in both theoretical and empirical literature.

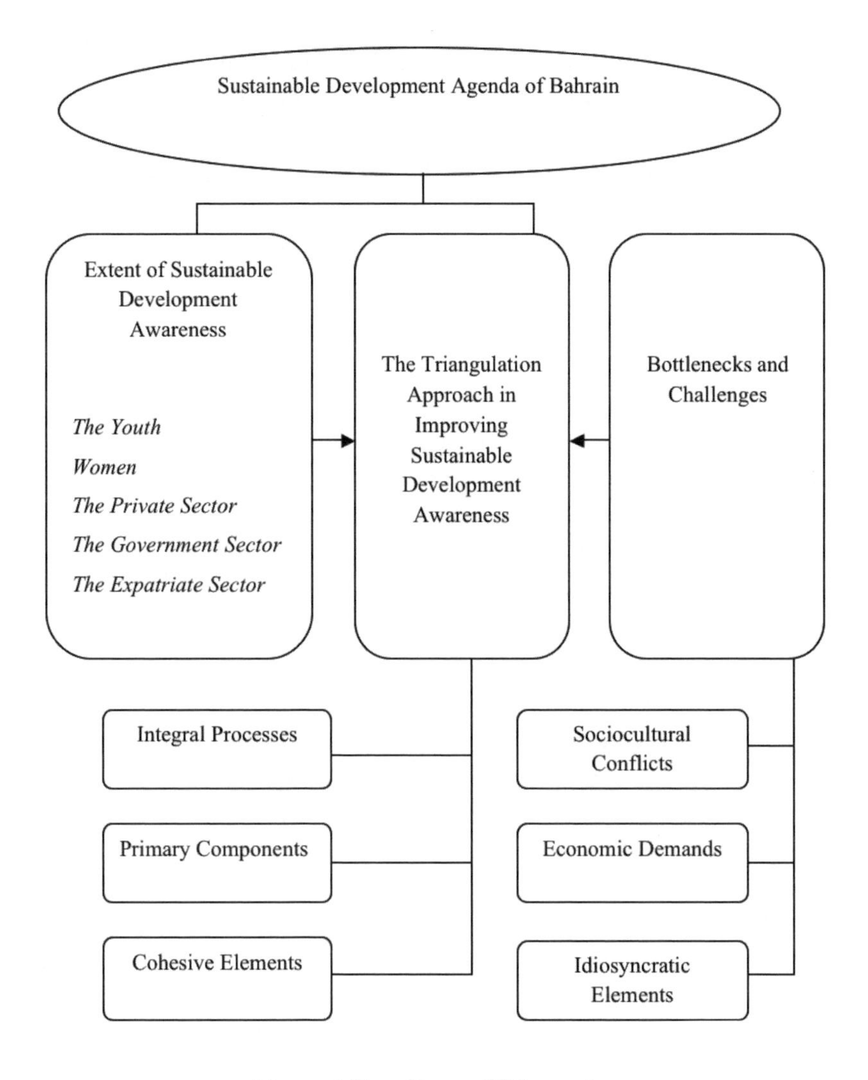

Figure 1 **Paradigm of Discussion**

The second section of the chapter builds a discourse that centers on the specific actions that are needed to be undertaken in order to address the current status of Bahrain in terms of its Sustainable Development Awareness (SDA). Using the Triangulation Approach (TA), the aforesaid measures are tackled considerably. In the exposition of the said approach, the integral processes, primary components, and cohesive elements are described in detail. Primarily, the planning, implementation, and assessment are the main components of the Triangulation Approach. The role of the youth, women, and the grassroots sectors in the promotion of Sustainable Development Awareness is given foremost significance under the TA. An integral part of the TA is the inclusion of the different social sectors in a process that is based on strong collaboration and partnerships.

As seen from the illustration in Figure 1, the last section of the chapter offers a discussion on the anticipated challenges in the future utilization of the aforesaid approach. This includes the sociocultural conflicts, economic demands, and idiosyncratic elements. Diverse sectoral implications on the advancement of Sustainable Development Awareness among various stakeholders of the Bahrain community are clarified in the last section of the chapter. Identifying and discussing the anticipated challenges is essential in the successful implementation of the proposed approach in the Kingdom of Bahrain.

2. Literature Review

In this section, the elements of sustainable development are clarified since one of the objectives of this endeavor is to determine the level of sustainable development awareness among the various sectors in Bahrain. Reviewing the constructs of awareness, consciousness, perception, and similar concepts in relation to sustainable development also provides strong cohesion and support in the development of a sound and logical procedural methodology.

On the Elements of Sustainable Development. An important phase in the formulation of a relevant framework for sustainable development awareness in Bahrain is to plough the misconceptions about sustainable development. This can be achieved by breaking down the theoretical and conceptual constructs of the focal point of inquiry by examining its elements, frameworks, and models. Over the years, there has been a plethora of constructs associated to sustainable development. In the model proposed by Mathey (2014), the financial, governance, social, health and safety, and environment aspects were identified to be key elements of sustainable development wherein the process of aligning financial and sustainability targets, reporting transparency in all phases of operations, ensuring environmental safety and conservation, promoting employee development and well-being, and fostering community and client benefits are all indispensible targets.

One of the most recognizable models of sustainable development is the Three Pillars Model, which acknowledges three important elements: society, environment, and economy. It emphasizes environmental conservation, economic growth, and social equity dimensions (Centre for Environment Education, 2007). The Egg of Sustainability Model, which was introduced by the International Union for the Conservation of Nature, tackles another perspective of the aforesaid model by accentuating the value of environment or ecosystem. In other words, it embraces the principle that environmental conservation precedes sustainable development (Guijt & Moiseev, 2001). An expansion to the aforesaid two models of sustainable development is the Prism Model of Sustainability, which adds the concept of institution as another dimension to the three original elements -- economy, environment, and society. It is the institutional dimension that is concerned with the organization of society and the transactions between people. Hence, the aforesaid model places premium on the interaction of the four dimensions and puts forward the principle that there is constant overlapping among these elements (Stenberg, 2001). An important abstraction from these three models is the pivotal role of society, environment, and economy.

Government plans for sustainability have always been a valuable resource in examining the common elements of sustainable development. The Sustainable Development Plan of Syracuse Onondaga County (2011) in New York is important to review in this chapter since it pinpoints several key elements in the implementation of the sustainability agenda. It highlights the area of land and transportation use, which evaluates relationships between the provision of transportation infrastructure and land use development patterns; the area of buildings and neighborhoods, which examines urban form, sustainable development patterns and development regulations; the area of rural communities and open space, which examines the impact of development in rural areas where agriculture, scenic vistas and natural resources are highly valued; the area of government finance, which explores how current development patterns affect the ability of local governments to pay for and provide the high quality of life expected by their resident; the area of water resources, which considers how the resulting development patterns impact the quality of our water; the area of intermunicipal planning, which identifies the various levels of planning that occur and examines the current extent of intermunicipal cooperation between municipalities and government agencies; the area of energy, which examines the impact of current development trends on energy consumption and greenhouse gas emissions and explores the viability of various renewable energy sources; the area of economy, which examines the impact that local government structure has on the regional economy and evaluates the role that quality of life plays in attracting and retaining businesses and employees; and the area of livability and society, which examines the relationships between development patterns, quality of life, and social equity (Sustainable Development Plan of Syracuse Onondaga County, 2011). From the aforesaid elements in the sustainable development plan, it can be deduced that society, economy, and environment are central and are given important emphasis.

Some frameworks in sustainability integrate the use of ICT. This approach is highly reflected in the Smart Sustainable Development Model proposed by the International Telecommunication Union. In this model, ICT linkages are put forward to facilitate community resilience, assurance, service delivery, enabling environment, human development, humanitarian, and statistical processes which are all essential elements in sustainable development. It is based on the objective of linking "rural telecommunications/ ICT development for general communications, business, education, health, and banking to disaster risk reduction and disaster management initiatives… [in order to] ensure the optimal use of technology, avoid duplication and lower the level of investment into ICT..." (International Telecommunication Union, p. 9, 2012). It can be subsumed from this paradigm that social, economic, and environmental aspects of human activities are considered to be significant aspects of sustainability. Social and economic processes must be based on equity; economic and environmental processes must be based on viability; and social and environmental processes must be bearable in order to promote the agenda of sustainability.

On the Constructs of Sustainable Development Awareness. It is evident that society, economy, and environment are the core elements of sustainable development based on the aforementioned conceptual literature. While reviewing these elements is essential in establishing common grounds for sustainable development, an equally important aspect is throwing light on the concept of sustainable development awareness in order to propose a relevant improvement scheme in the context of Bahrain. Awareness is defined as "the state or ability to perceive, to feel, or to be conscious of events, objects, or sensory patterns. In this level of consciousness, sense data can be confirmed by an observer without necessarily implying understanding" (Wikipedia, para 1, 2014). This concept is often associated with consciousness, which is viewed as "the quality or state of self-awareness, or, of being aware of an external object or something within oneself" (Van Gulick, sec 2, 2004) The presence of external stimulus and the ability to recognize its presence are common elements in these two conceptualizations. Hence, a superficial view of sustainable development awareness is the ability of an individual to perceive the mechanisms and strategies for sustainable development. Nonetheless, this paradigm proves to be limited and needs further conceptual exploration.

Sustainable development awareness is a variable that has received attention only in recent years. An interesting point to highlight is the common observation that most of these attempts in clarifying the concept of sustainable development awareness only tackled the periphery. For example, Razman *et al* (p. 194, 2012) claimed that environmental awareness precedes sustainable development by building their argument on the principle of tranboundary liability, which is a "response to the need of individuals to protect their rights and interests…in order to meet the needs of the present without compromising the ability of future generations to meet their own needs." Environmental awareness, as argued in the aforesaid proposition, is essential in advancing sustainable development. Upon a careful examination of this argument, it can be deduced that Razman *et al* did not really explore the concept of sustainable development awareness since the focus of their study was only to highlight environmental awareness.

Similar to the attempts initiated by Razman *et al* (2012), the work of Shastri (2005) also applied awareness as one of the variables. He argued that "sustainable development depends upon participation by the people, and their awareness of the environmental effects of their actions" (Shastri, para 1, 2005). Crotty and Hall (2012) explored the same objectives by utilizing data from approximately 100 surveys with companies and Non-Government Organizations in order to put forward the relationship of environmental awareness and sustainable development in Russian society. Similarly, the paper of Ugurlu and Aladag

(2004) argues that environmental educational process is one of the most important tools in appreciating the essence of sustainable development. These three empirical studies on sustainability only focused on the periphery of the sustainable development awareness construct – a feature that poses certain applicability limitations to the present inquiry.

On a different note, the work of Evelyn and Tyav (2012) is a concrete example of a study that tackles the heart of sustainable development awareness variable. In their said study, the environmental pollution in Nigeria was described and analyzed as bases for recommending a sustainable development awareness mechanism. The relevance of the said paper can be drawn from its recommendation to inculcate awareness on the stakeholders and modify their attitude and perceptions for effective environmental and resource management. This proposition finds support in the report of Dawe, Jucker & Martin (p. 4, 2005). about the UK's new sustainable development strategy, which highlights "the role that education can play in both raising awareness among young people about sustainable development and giving them the skills to put sustainable development into practice." The said proposal for sustainable development education has provisions for specific teaching methods, teaching approaches, curricular content, emerging good practices, and barriers and solutions for its effective implementation.

In one empirical research, junior high school students' awareness of climate change and sustainable development in the central region of Ghana was assessed using a descriptive survey design and a purposive sampling technique. An important finding in this study is the low level of sustainable development awareness among the youth as manifested in the responses of the junior high school students (Owolabi, Gyimah, & Amponsah, 2012). This study of Owolabi *et al* (2012) holds primary relevance to the present inquiry since one of the objectives of this chapter is also to determine the extent of sustainable development awareness of the various categories of stakeholders in Bahrain.

Choi's (2006) attempt to provide a description of the different aspects or characteristics of sustainable development awareness by focusing on human behaviors is worthy to note. In his paper, he enumerated these characteristics: appearance of the voice for the rights of future generations, advancement of environmental discussions, popularization of the debates on the relationships between the environment and economic growth, appearance of governmental or non-governmental organizations, presence of the community-level salient event, influx of global environmental agreements, emergence of environmental issues, transboundary phenomena regardless of scales or levels, emergence of taking initiatives about environmental protection, appearance of new governance in implementing development agenda, emergence of new perspectives, presence of extreme and innovative reactions to environmental problems, appearance of new actors, appearance of request for fundamental change, occurrence of local-type globalization, appearance of unprecedented trend of multidisciplinary studies in academic circles, awareness of reaching a boundary or limit, and awareness of the existing institution's inability (Choi, 2006). This list may seem to be exhaustive but the constructs of sustainable development awareness can be summarized into certain categories depending on the context and environment of its application.

In Ireland, sustainable development awareness is operationalized in its effort to educate the stakeholders. Ireland's Framework for Sustainable Development specifically provides that "education for sustainable development is crucial in strengthening the capacity of individuals, communities, businesses and governments to make judgments and decisions in its favor. It needs to be embedded at every level of the education system" (Government of Ireland, sec 3, 2014). This strategy is based on the assumption that public awareness can improve the level of sustainability of a community, which remains to be a challenge to the law-making bodies. An interesting question is how to maintain this level of awareness through the dynamic interaction of time and place. In the work of Johansson, Rex, Nyström, Wedel, Stahre, & Söderberg (para 1, 2012), the role of research in sustaining this awareness was emphasized. They argued that "awareness of sustainability is the key to create sustainable products, and that this awareness begins already at research level."

The efforts to increase sustainable development awareness through education stems from the mandates of the United Nations particularly in its Decade of Education for Sustainable Development (DESD) program. This program aims to "integrate the values inherent in sustainable development into all aspects of learning to encourage changes in behavior that allow for a more sustainable and just society for all," (United Nations Scientific and Cultural Organization, sec 1, 2005) and it particularly delegates the United Nations Scientific and Cultural Organization (UNESCO) to "provide recommendations for governments on how to promote and improve the integration of education for sustainable development in their respective educational strategies and action plans at the appropriate level" (United Nations Scientific and Cultural Organization, sec 1, 2005).

3. Methodology

Design and Approach. With the given objectives at the onset of the project, it was imperative to use a normative descriptive approach. This design was justified by the nature of the task to describe the extent of sustainable development awareness of the various stakeholders of Bahrain. The proposed Triangulation Approach was also described in terms of its integral processes, primary components, and cohesive elements in its implementation, aiming to enhance Sustainable Development Awareness (SDA) of the region. The need to document the issues and anticipated challenges also rationalizes the utilization of this design.

Data Sources. Both primary and secondary sources of data were applied in this inquiry. Foremost, the use of primary sources of information through an informal focus group discussion was deemed important in addressing the objective of determining the extent of sustainable development awareness of various stakeholders in Bahrain. The respondents for this focus group discussion included the youth, women, private sector, government sector, and expatriate sector of the Bahrain community.

Secondary sources of information were considered sufficient to provide light on the second objective of the chapter which is to describe the implementation of the proposed Triangulation Approach. Theoretical, conceptual, and empirical models of sustainable development found in various literatures were analyzed in order to provide baseline structure for the proposed approach in advancing the sustainable development awareness of the stakeholders. The data sources used as bases in identifying the anticipated challenges and issues associated with the implementation of Triangulaton Approach also consisted of the available conceptual and empirical literatures which are considered as secondary sources.

Procedure and Techniques. Two mutually distinct procedures were applied in this endeavor in order to address the project objectives. An ***informal focus group discussion*** (FGD) technique was specifically applied to determine the extent of stakeholders' sustainable development awareness. The Research and Policy in Development (para 1, 2009) described this procedure as "a good way to gather together people from similar backgrounds or experiences to discuss a specific topic of interest. The group of participants is guided by a moderator ... who introduces topics for discussion and helps the group to participate in a ... discussion." From this contention, it is logical to argue that the stakeholders' sustainable development awareness can be elicited effectively through the application of focus group discussion. Elliot & Associates (para 2, 2005) explained the positive features of this procedure by stating that "focus groups can reveal a wealth of detailed information and deep insight. When well executed, a focus group creates an accepting environment that puts participants at ease allowing then to thoughtfully answer questions in their own words..." It must be noted that the present inquiry modified some of the features of the standard focus group discussion in order to accommodate other categories of respondents. While the standard focus group discussion allows participants to be in a homogeneous group, the present inquiry included respondents from different backgrounds and experiences to probe the generalizability of results.

The ***systems analysis*** technique was also utilized to address the second project objective. The Web Dictionary of Cybernautics and Systems (para 1, 2007) delineates system analysis as an "explicit formal inquiry carried out to help someone (referred to as the decision maker) identify a better course of action and make a better decision than he might otherwise have made." Its applicability in this endeavor is based on the premise that the level of stakeholders' Sustainable Development Awareness was determined first before formulating a relevant framework to address the former. Thus, in the formulation of the guidelines and mechanics of the Triangulation Approach, it was needful to analyze first the present situation, context, and status of sustainable development awareness of the sectoral stakeholders.

Elements of the ***meta-analysis*** technique were also applied to address the second and third project objectives. Greenland and O' Rourke (2008) explained that meta-analysis includes contrasting and combining previous empirical findings in order to identify important patterns, relationships, models, and interactions of the focal variables that are under examination. Egger (para 2, 2007) also emphasized that "meta-analysis should be viewed as an observational study of the evidence. The steps involved are similar to any other research undertaking: formulation of the problem to be addressed, collection and analysis of the data, and reporting of the results." It is apparent that the principles of meta-analysis technique were highly applicable in this inquiry since examination and utilization of various theoretical and conceptual models of sustainable development awareness were the primary bases in the development of the Triangulation Approach and in the identification of the possible challenges and issues in its implementation.

4. Discussion

4.1. Status of Sustainable Development Awareness in Bahrain

The Youth and Women. Based on the informal focus group discussion, the youth and women sector revealed to have low level of sustainable development awareness. As reflected in Table 1, this group of respondents obtained a relatively low level of basic understanding that economy, society, and environment are the three core elements of sustainable development- elements that are integrated in the Three Pillars Model, Egg of Sustainability Model, and Prism Model of Sustainable Development. They also obtained a low level of awareness in terms of the characteristics proposed by Choi and in terms of their conscious knowledge of the practices and strategies in sustainable development that are officially implemented and mandated in the Kingdom of Bahrain. During the entire duration of the discussion, a few of them raised some points regarding the issues and programs in sustainable development that are manifest in the country.

This low level of awareness on sustainable development among the youth is a local concern that needs to be addressed. The role of youth in advancing sustainable development programs is crucial. In fact, this is acknowledged by the United Nations Environment Program (UNEP) (para 3, 2014), wherein it proposes that the youth sector should be involved "...actively in all relevant levels of decision-making processes because it affects their lives today and has implications for their futures. In addition to their intellectual contribution and their ability to mobilize support, they bring unique perspectives...." Educating the youth sector is the initial step in realizing this goal. Corcoran and Osano (2014) placed great emphasis on the task of inculcating sustainable development awareness among the youth by arguing that they can be potential effective catalysts of the sustainability agenda through the supervision of different governing bodies.

Table 1 Sustainable Development Awareness of Bahrain Stakeholders

Elements and Models of Sustainable Development Awareness as Indicators	Respondent Categories	Extent/ Level of Sustainable Development Awareness
Basic understanding that economy, society, and environment are the three core elements of sustainable development (elements integrated in the Three Pillars Model, Egg of Sustainability Model, and Prism Model)	Youth and Women	Low
	Government	High
	Private and Expatriate	Low
Awareness on the important role of environmental conservation, management, and protection in the overall advancement of sustainable development	Youth and Women	Moderate
	Government	High
	Private and Expatriate	Moderate
Reflective of at least five characteristics of sustainable development awareness proposed by Choi	Youth and Women	Low
	Government	High
	Private and Expatriate	Low
Conscious knowledge of the practices and strategies in sustainable development that are officially implemented and mandated in the Kingdom of Bahrain	Youth and Women	Low
	Government	High
	Private and Expatriate	Low
Practical understanding on the application of the basic elements of sustainable development in everyday official and personal activities.	Youth and Women	Moderate
	Government	High
	Private and Expatriate	Moderate

It can be gleaned from the table that a moderate level of awareness was obtained by the youth and women sector in terms of their consciousness on the important role of environmental conservation, management, and protection in the overall advancement of sustainable development. In the same manner, the youth and women sector obtained a moderate level of practical understanding on the application of the basic elements of sustainable development in everyday official and personal activities. These findings pose an interesting implication for future inquiries in the sense that a variation of the stakeholders' extent of awareness was demonstrated between these two indicators and the previous three indicators of sustainable development awareness. This can be attributed to the nature of constructs presented in the three earlier indicators of sustainable development awareness. As it can be gleaned from these constructs, the three earlier indicators (wherein the respondents obtained a low level of awareness) are constructs that have theoretical and abstract orientation compared to the two indicators (wherein the respondents obtained a moderate level of awareness). To put simply, respondents tend to have higher awareness and understanding on practical concepts such as their practical understanding on the application of the basic elements of sustainable development in everyday official and personal activities than abstract or theoretical concepts such as the elements of sustainability.

Regardless of the variation observed based on the two types of indicators for sustainable development awareness, it is important to consider that the women sector is equally important as the youth sector. The Organization for Economic Cooperation and Development (OECD) (p. 4, 2008) acknowledges that "better use of the world's female population could increase economic growth, reduce poverty, enhance societal well-being, and help ensure sustainable development in all countries. Closing the gender gap depends on enlightened government policies which take gender dimensions into account." Policy makers may consider strategies in enhancing their level of awareness on sustainable development in order to efficaciously achieve the agenda for sustainability.

The Government Sector. A relatively high level of sustainable development awareness was noted in the government sector. This is reflected in all the indicators of sustainable development awareness, which specifically includes their basic understanding that economy, society, and environment are the three core elements of sustainable development; their awareness on the important role of environmental conservation, management, and protection in the overall advancement of sustainable development; their possession of at least five characteristics of sustainable development awareness proposed by Choi; their conscious knowledge of the practices and strategies in sustainable development that are officially implemented and mandated in the Kingdom of Bahrain; and their practical understanding on the application of the basic elements of sustainable development in everyday official and personal activities.

The relatively high level of sustainable development awareness among the respondents representing the different agencies of the government sector in Bahrain is not surprising. This is because of the government initiatives to foster sustainability in the region. The e-government portal of Bahrain reports that the country "has made great efforts in ensuring the sustainability of the development process in multiple areas - reflected in the Constitution of the Kingdom" (Kingdom of Bahrain E-Government Portal, para 1, 2014). This is also stipulated in the Bahrain Economic Vision 2030, the National Strategic Master Plan 2030, and Bahrain's National Action Charter, which have been accepted by the stakeholders to promote environmental protection and the equilibrium of social and economic developments. As employees in the government sector, the respondents of the focus group discussion are expected by the state not only to be aware but also to be highly cognizant of the programs of the government in advancing sustainablity. During the discussion, the participants from the government sector have comprehensive knowledge and understanding of these initiatives. Some sustainable development programs that were explicitly mentioned by the participants included the Bahrain

National Environment Strategy, the Millennium Development Goals Report, the Supreme Council for Environment, and some environment protection decrees and decisions.

Stakeholders from the government sector were also aware of other efforts of the state to foster sustainable development. Although the discussion was not comprehensive, they explicitly mentioned about their awareness on the environmental laws of the country, the Wildlife Act, Law of Marine Wealth, the National Strategy for the Environment, and the Protection of the Environment Institutionalization program. Some respondents during the informal discussion also raised some points on the government's efforts to protect the environment which are translated into the areas of energy, water, air quality, biodiversity, land use and desertification, coastal areas, chemical substances, and waste. This relatively high level of awareness among the respondents from the government sector can be attributed to their duties and obligations to know the legislations of the country and to be highly cognizant with the provisions inherent in their functions in order to effectively perform their roles. Ignorance on these laws and provisions would be tantamount to inefficiency.

The Private and Expatriate Sector. Those who belong to the private and expatriate sector obtained a relatively low level of sustainable development awareness in terms of their basic understanding that economy, society, and environment are the three core elements of sustainable development; their possession of at least five characteristics of sustainable development awareness proposed by Choi (2006); and their conscious knowledge of the practices and strategies in sustainable development that are officially implemented and mandated in the Kingdom of Bahrain. However, the aforesaid group obtained a moderate level of awareness in terms of their practical understanding on the application of the basic elements of sustainable development in everyday official and personal activities; and in their awareness on the important role of environmental conservation, management, and protection in the overall advancement of sustainable development.

Unlike the government sector, respondents from the private and expatriate sector raised minor points of discussion during the FGD. Only a few of them were able to give inputs regarding the environmental laws of the country, the Wildlife Act, Law of Marine Wealth, the National Strategy for the Environment, and the Protection of the Environment Institutionalization program. Consequently, they were able to point out minor suggestions and recommendations regarding the government's efforts to protect the environment which are translated into the areas of energy, water, air quality, biodiversity, land use and desertification, coastal areas, chemical substances, and waste.

The results indicate that the private and expatriate sector has almost similar level of awareness with the youth and women sector. This low to moderate level of awareness is also a concern that needs to be addressed because the private and expatriate sector comprises a majority of the population of the country. Their important role in advancing sustainablity, given these demographic characteristics and contexts, is acknowledged in several literatures. In one of the reports for the United Nations (para 2, 2012), it was highlighted that private companies are now "realizing the growing significance and urgency of global environmental, social and economic challenges. Using science and technology to stimulate innovation and investment for green growth, the private sector is contributing substantially to sustainable development." However, this is not easy to achieve due to the complexity of this process. Garcia (p. 140, 2014), for example, explained how private sectors can contribute to sustainable development by accentuating that this process will involve "...close partnership between government, business, and financial institutions. Through this partnership, all actors can encourage the adoption of eco-efficient principles which balance environmental, social, and economic factors for the good of society."

Given the varying level of sustainable development awareness among the different stakeholders of Bahrain, it is deemed needful to formulate a framework that would aim to increase the current extent of awareness on sustainability. This is tackled in the next section of this chapter, wherein the Triangulation Approach is proposed.

4.2. Reinforcing Sustainable Development Awareness through Triangulation Approach

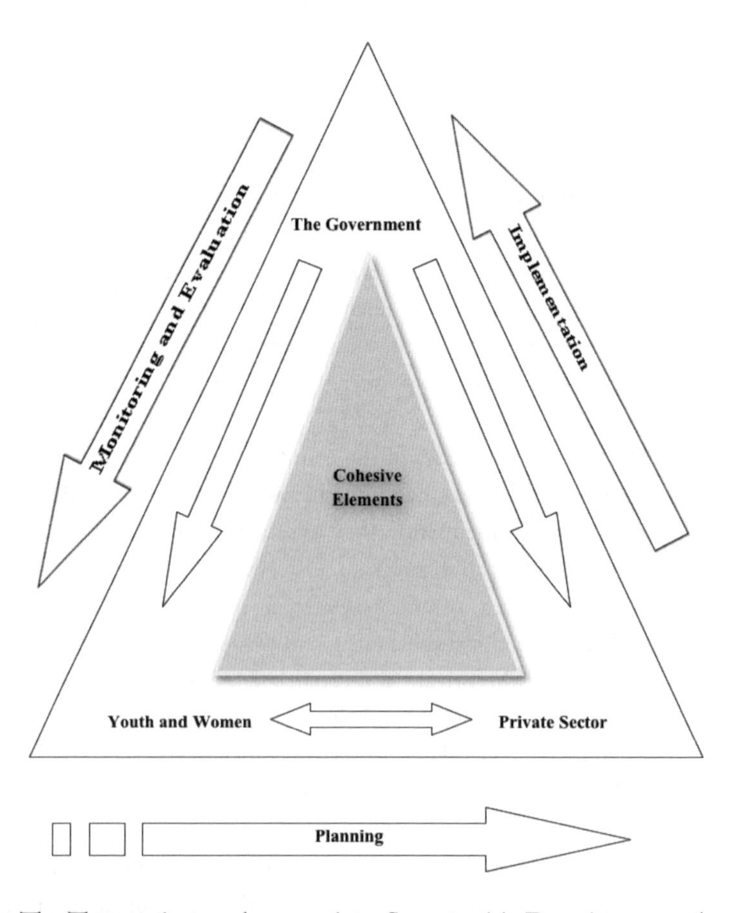

Figure 2 The Triangulation Approach in Sustainable Development Awareness

Integral Processes. The integral processes in the Triangulation Approach include the planning, implementation, and monitoring or evaluation. As depicted in the schema in Figure 2, these integral processes are reflected in the outermost layer of the triangle. The baseline process is planning, which is depicted in an arrow at the base of the triangle. This process constitutes the initial phase of implementing the framework for sustainable development awareness in the Kingdom of Bahrain. Without proper planning, promulgation of other succeeding processes would not be effective.

In the planning phase, the government and policy makers primarily considers the different variables that might affect the implementation of the sustainable development awareness program. It analyzes the past and present situation and conditions in order to carry out the goals of sustainable development awareness. Formulation of the vision, mission, goals, and objectives of the overall government framework is vital in operationalizing the targets of the plan.

The second integral process of the Triangulation Approach in increasing the sustainable development awareness of the stakeholders in Bahrain is the implementation phase. This is considered to be the formal phase of carrying out the objectives and goals established during the planning stage. As gleaned from the schema, this is represented by an upward arrow because of the nature of this stage to be carried out with certain targets and timeframes. Coordination among the various departments of the Kingdom is an essential feature of this approach.

Monitoring and evaluation forms the downward slope among the integral processes as indicated by the descending arrow. This is due to the nature of this process to be at work when the implementation stage is done. At this stage, the policy makers and the stakeholders are required to determine the extent of effectiveness of the various strategies implemented by the government to promote and increase the sustainable development awareness in the region. Monitoring whether a particular strategy is effective or not during and after the implementation gives advantageous insights not only to the general stakeholders but most primarily to the policy makers since they would be informed of the facilitative or debilitating effect of the policies on sustainable development awareness that they have established. Evaluation at the end of an implementation cycle would also inform the state and other implementing bodies whether a specific strategy is to be removed or retained.

The Triangulation Approach paradigm also indicates that the integral processes are presented in a manner that reflects a triangular cycle. It is a triangular cycle since the movement of the arrows in the outer triangle, as depicted in Figure 2, does not cease or terminate. After the monitoring and evaluation stage, the results of sustainable development awareness program effectiveness has to be communicated to the policy makers. These results are valuable insights and bases in the development and formulation of the next planning stage for sustainable development awareness implementation cycle.

Primary Components. It can be gleaned from the paradigm that the primary components in the implementation of the Triangular Approach are composed of the government sector, the youth and women sector, and the private and expatriate sector. Figure 2 shows that at the apex of the triangle is the seat of the government sector. This signifies that all policies and mandates are formulated by the state since Bahrain is a constitutional monarchy and all political decisions originate from the Parliament or a bicameral National Assembly consisting of the Shura Council and the Council of Representatives with the leadership and influence of the head of state, King Hamad bin Isa Al-Khalifa (Bahrain, 2010). In other words, a top-down approach is the key to this triangulation model, wherein the government sector drafts and implements the programs for sustainable development awareness for its entire jurisdiction.

At the base of the triangle are two important components of the Triangulation Approach. These are the youth and women sector and the private and expatriate sector. Both these sectors have equal bearing in the proposed effort to improve the sustainable development awareness of the nation. Based on the said illustration, it is assumed that mandates, provisions, and programs issued and implemented by the government must be complied by these sectors at the base of the triangle. While it is true that these base sectors are recipient of the mandates and provisions for sustainable development awareness as dictated by the nature of Bahrain's government, it does not imply that suggestions and inputs for policy formulation may not come from these sectors. The schema indicates that these sectors are the primary targets of the government in its goal to promote and increase sustainable development awareness. The two descending arrows pointing to these sectors suggest that the government's efforts to increase sustainablity awareness have to be directed to the sectors that have low to moderate level. The double-headed arrow between the youth and women sector and the private and expatriate sector depicted in the model suggests that there exists an overlapping interaction between these sectors. This interaction is based on the assumption that the youth and women sector is also a component of the private and expatriate sector and vice-versa. The bilateral coordination and collaboration between these two sectors is also depicted by the double-headed arrow.

Cohesive Elements. At the core of the Triangulation Approach are the cohesive elements, which refer to the aspects that facilitate the effective implementation of the government's effort to increase sustainable development awareness. One of the cohesive elements is the integration of Information Communication Technology (ICT) in the coordinated efforts to foster SDA. At present, the use of ICT is unparalleled in all sectors of the Kingdom. Its e-government initiatives are manifest in the delivery of basic social services to the civil society such as in the area of education, health and safety, commerce, environment, and other areas of governance. By maximizing and incorporating the potentials of ICT and its present application in the Kingdom of Bahrain, sustainable development awareness can be bolstered to greater heights.

Another form of ICT integration is the utilization of social media. This practice has proven to be pervasive in the recent years especially in the youth sector. These social media tools encourage collaborative projects such as the Wikipedia, promote blogs and microblogs such as Twitter, offer social news networking sites such as Digg and Leakernet, create content communities such as YouTube and DailyMotion, and cater to social networking such as Facebook and My Space. The potency of these social media in raising awareness on sustainable development in Bahrain is based on the precepts that these applications create a significantly interactive avenue for individuals and organizations to discuss and share ideas. In the report of Media Week (2014), the role of social media is acknowledged in the promotion and widespread realization of sustainable development.

Strengthening research and development activities forms another cohesive element in the Triangulation Approach of raising the sustainable development awareness in Bahrain. Developing a research culture that would regularly tackle the issues and concerns in sustainable development of the country is imperative in increasing the consciousness of the stakeholders towards this agenda. In building a research and development culture on sustainable development, scaffolding from other sectors is of paramount consideration since absence of these technical and administrative support would create ambiguous trajectories on the primary purpose of conducting empirical studies. By encouraging small-scale to nationwide cross-sectional and longitudinal researches on sustainable development, stakeholders' awareness on the issues related to sustainability would be enhanced. Financial support and grants from the government can create capability mechanisms for various groups of researchers on sustainability. International linkages and coordination with experts on this field may strengthen the pathway for continued growth of researches and investigations. Universities and other educational institutions can be appointed as part of a national committee that will oversee the implementation of researches in the country since these institutions have the mechanisms and thrusts to spearhead research activities with other ministries and experts in the Kingdom.

Finally, mainstreaming sustainable development activities and awareness programs to the smallest units of Bahrain community comprises another pivotal cohesive element in the goal of advancing Sustainable Development Awareness (SDA). This means that the national agenda, provisions, and policies for sustainable development are realized and translated into individual families and workplaces. This includes the government's efforts to protect the environment which are operationalized into the areas of energy, water, air quality, biodiversity, land use and desertification, coastal areas, chemical substances, and waste.

Mainstreaming implies that the aforesaid efforts and programs of the Kingdom are accepted and carried out even by the smallest unit of society. Different offices and institutions are also encouraged to harmonize their standards of implementation with the national policies for sustainable development. Hence, it can be argued that mainstreaming the goals of sustainable development is a double-edged sword. First, it facilitates the effective implementation of the sustainability agenda by translating the national goals and policies from the smallest unit of society to the highest echelons of the government. Second, it implicitly inculcates consciousness, awareness, and a deeper understanding of sustainable development to the various stakeholders of the Kingdom because the national policies are not only understood by the officials and appointed agents of the government ministries but are also assimilated to the individual families and units of society. In order to achieve the mainstreaming scheme effectively, careful coordination and collaboration between government agencies and ministries is needed with the integration of ICT and the enforcement of sustainable development research program.

4.3. Bottlenecks and Challenges

Similar to other models and frameworks, the Triangulation Approach is not free from any challenges and issues in the course of its implementation. Identification of these bottlenecks is an important process in its effective promulgation in the context of the Kingdom of Bahrain. These issues are categorized as sociocultural conflicts, economic demands, and idiosyncratic elements.

Sociocultural Conflicts. Efforts in raising Sustainable Development Awareness (SDA) may be hindered by the culture, traditions, and beliefs of the community. Svensson's (p. 369, 2012) empirical work, for example, indicated that "there is a gap between what households perceive as ideologically correct behavior and what they actually do... Socio-cultural dispositions, material culture and collective action need to be included in future strategies for creating more sustainable lifestyles." Implicit from the said argument is the influence of sociocultural factors in the assimilation of sustainable development lifestyle. The load of society and culture plays a crucial role in the successful implementation of any national goals including sustainability agenda.

Furusawa (2012) also acknowledges the influence of sociocultural elements in sustainability and the potential issues it may pose on the latter. Vitanen and Saarinen (p. 226, 2012) support this claim by stating that "the best means of integrating socio-cultural dimensions into the study and politics of sustainable development are through the ethnographic analysis of cultural models and patterns of behavior...the traditional cultural models and patterns of behavior are reproduced in interaction with new non-native actors..." It can be put forward from these assumptions that when the sociocultural paradigm of a particular community is not aligned with the goals of sustainable development, meeting this objective would be hindered. This can also happen when there is sociocultural unrest or instability such as internal conflicts among its stakeholders, insurgencies, or threats brought about by terrorism. In the Kingdom of Bahrain, there have been documented cases of these sociocultural problems, which were successfully resolved by its government through the assistance of its allies and international communities. Although these issues have been pacified by the state, their possible recurrence should not be taken for granted. As such, it is imperative to devise mechanisms and intervention to address this possible variable that can affect the implementation of the Triangulation Approach.

Economic Demands. The Kingdom of Bahrain has been noted to be the freest economy in the Middle East and North Africa according to the 2011 Index of Economic Freedom (Central Intelligence Agency, 2014) and it is also the tenth freest economy in the world. Bahrain was ranked in the 44th place together with 7 other countries, and it is also considered as a country with high economy by the World Bank (2013). This economic demand to maintain or improve Bahrain's economic status and prestige in the world's economy poses a threat to the implementation of the goals of the Triangulation Approach, albeit this economic demand and pressure is inherently for the welfare of the nation. What makes this economic demand debilitating is when the goals of sustainability are compromised.

In the study of Chang, Yeh, & Chen (para 1, 2013), it was stressed that the "increase in the carbon footprint resulting from economic growth cannot be counterbalanced by technological advances in environmental protection at different stages of economic development. Furthermore, international trade, industrial structure and energy demands have significant effects on the carbon footprint." In the context of Bahrain, the pressure for economic development is obstinate, which also holds ground for the same increase in carbon footprint. Forstater (p. 388, 2003), in support to this claim, contends that environmental ruin "in the form of unsustainable rates of natural resource depletion and excessive pollution of land, air, and water is characteristic of modern capitalist economies. Humanity now faces significant challenges in the form of both local ecological crises and global environmental problems." Given the significant influence of the economic demands in sustainable development implementation, it can also be deduced that Sustainable Development Awareness (SDA) would also be compromised under the effect of the former. The private sector such as multinational companies and corporation are expected to be the direct targets of this economic demand. Thus, it is a challenge among policy makers to ensure that these companies become increasingly aware and highly conscious of the national goals for sustainability.

Idiosyncratic Elements. Factors inherent in every individual stakeholder may indirectly affect the implementation of the Triangulation Approach. This includes the personality, attitude, and other idiosyncratic attributes of each individual. Once an individual has a negative attitude towards the political system of a country, his understanding and awareness on the issues related to sustainability policies of the nation can also be affected negatively. The United Nations through its United Nations Decade for Education for Sustainable Development program articulates the effects of individual attitudes in the implementation of sustainable development. The said program calls for a behavior modification towards sustainable development (Creech, McDonald, & Kahlke, 2011).

The importance of idiosyncratic factors such as individual attitudes is also emphasized by Leiserowitz, Kates, & Parris (p. 22, 2005) in their assumption that "...global sustainability—meeting human needs and reducing hunger and poverty while maintaining the life-support systems of the planet—will require changes in human values, attitudes, and behaviors." Explicit in this conjecture is the centrality of human values, attitudes, and behaviors in every national policy. This has also verity and applicability in the proposed Triangulation Approach. The government and other implementing bodies in the Kingdom should consider the values, attitudes, and behaviors of the citizens and stakeholders when it comes to the goal of raising sustainable development awareness and understanding. Aligning state policies and programs with the values, attitudes, and behaviors of the citizens would maximize the potentials for program effectiveness and efficiency rather than channel the state resources into waste.

5. Conclusions

This endeavor attempted to describe the extent of sustainable development awareness of the different stakeholders in Bahrain. It also utilized these inputs in developing a framework with the goal of advancing the awareness on sustainability. This is reflected in the proposed Triangulation Approach. The results of the informal focus group discussion revealed the need to raise the consciousness and understanding of the stakeholders in terms of the issues that confront the sustainability agenda of the nation. Creating other instruments in measuring the level of sustainable development awareness would contribute to the generalizability of the current results obtained from the qualitative method of focus group discussion. Utilization of a standardized research instrument and application of canonical analysis can enhance the reliability of the current findings. Nonetheless, the variation of the responses and extent of understanding of the three different groups of stakeholders in Bahrain suggests that there is a need to leverage the national policy on sustainable development awareness and education.

Effective implementation of sustainable development goals in Bahrain can be achieved only when the stakeholders are highly aware and conscious of these efforts and national policies. This is the core rationale for proposing the Triangulation Approach, which is a framework that addresses the issue of raising the level of Sustainable Development Awareness (SDA) of the different sectors in the Kingdom. The integral processes include planning, actual implementation, and monitoring and evaluation. The overlapping of these integral processes implies that there is always a room for flexibility along the continuum of promulgation. Permeability has to be considered in all stages of its implementation in order to allow constant redirection of the policies for sustainable development awareness based on the current needs of the community. The triangular cycle suggests that the flow of implementation starts with careful planning and definition of goals and objectives which are subjected to evaluation and monitoring along the course of program delivery. With the emphasis of monitoring and evaluation, policy makers should consider the creation of a committee that would regularly evaluate the effectiveness of program implementation in the country.

The form of governance in Bahrain implies for a top-down approach in terms of policy making. Hence, the accountability of its leaders and governing bodies has to be ensured and reinforced in order to carry out the goals of sustainable development effectively. On the contrary, this does not imply that the government can disregard the inputs and insights from the youth and women sector and the private sector. In fact, the Triangulation Approach indicates that these sectors have significant roles in ensuring that sustainability development goals are understood and operationalized.

At the heart of the Triangulation Approach are the cohesive elements of integrating ICT, developing a research culture on sustainable development, and the mainstreaming of the national policies on sustainable development to the individual stakeholders' personal and professional lives. Integration of ICT forms an important cohesive element in raising the level of Sustainable Development Awareness (SDA) in Bahrain because of its propensities to allow individuals to share and discuss issues in real time. Developing a research culture on sustainable development serves as an explicit attempt of the government to deepen the understanding of the various sectors on the issues of sustainability by encouraging them to engage in empirical studies. Mainstreaming the goals of sustainable development to the smallest units of Bahrain society would ensure alignment of personal and professional interests with the ultimate end of these national policies. Given the centrality of these three cohesive elements, it is imperative that government programs incorporate such aspects.

While the Triangulation Approach offers a promising framework in increasing the level of Sustainable Development Awareness (SDA) in the region, some anticipated variables, if not carefully controlled, can lead to the detriment of the former. This includes the threats of sociocultural conflicts, economic demands, and idiosyncratic dimensions. These bottlenecks and challenges cannot be completely eradicated in the milieu of the Kingdom since these have been part of its composition and structure. Nonetheless, these can be controlled and channelled into a vessel of opportunities for sustainable development awareness. The task of controlling the debilitating effects of these factors remains to be a challenge to the government sector and a recommendation for future inquiry and policy making.

References

Brittlebank, W. (2012). Bahrain's sustainable development projects. [online] Available at: <http://www.climateactionprogramme.org/news/bahrains_sustainable_development_ pro jects/> (Accessed 24 April 2014).

Brown, P. (2013). Sustainable development efforts mostly fail, research finds. *The Daily Climate*. [online] Available at: <http://www.dailyclimate.org/tdc-newsroom/2013/ 01/sustainable-development-goals> (Accessed 26 April 2014).

Central Intelligence Agency (2014). The world fact book: Bahrain. *Central Intelligence Agency*. [online] Available at: <https://www.cia.gov/library/publications/the-world-factbook/geos /ba.html#Econ> (Accessed 28 April 2014).

Centre for Environment Education. (2007). Sustainable development: an introduction. *Internship Series,* 1. [online] Available at: <http://www.sayen.org/Volume-I.pdf> (Accessed 28 April 2014).

Chang, D.S., L.T. Yeh, & Y. Chen. (2013). The effects of economic development, international trade, industrial structure and energy demands on sustainable development. *Sustainable Development*. [online] Available at: <http://onlinelibrary. wiley.com/doi/10.1002 /sd.1555/abstract> (Accessed 28 April 2014).

Choi, In Huck. (2006). Awareness of sustainable development: why did the Saemangeum Tideland Reclamation Project led to the first national controversy over sustainable development in South Korea. Master's Thesis. Texas A&M University.

Connors, S.R. (1998). Issues in energy and sustainable development. AGS Mapping Project White Paper – Energy." *EL*, 98-004. [online] Available at: <https://mitei.mit. edu/system/files/1998-mitenergylab-04-rp.pdf> (Accessed 5 May 2014).

Corcoran, P.B. & Philip M.O. (2014). Young people, education, and sustainable sevelopment: exploring principles, perspectives, and praxis. [online] Available at: <http://www.wageningenacademic. com/youngpeople> (Accessed 15 April 2014).

Crotty, J. & S. M. Hall. (2012). Environmental awareness and sustainable development in the Russian Federation. *Sust. Dev.* [online] Available at: <doi: 10.1002/sd.1542> (Accessed 15 April 2014).

Dawe, G., R. Jucker, & S. Martin. (2005). Sustainable development in higher education: current practice and future developments. [online] Available at: <http://thesite.eu/ sustdevinHEfinalreport.pdf> (Accessed 25 April 2014).

Dictionary of Cybernautics and Systems. (2007). System analysis. *Dictionary of Cybernautics and Systems.* [online] Available at: <http://pespmc1.vub.ac.be/ASC/SYSTEM_ ANALY.html> (Accessed 25 April 2014).

Egger, M., G.D. Smith, & A.N. Philips. (1997). Meta-analysis: principles and procedures. [online] Available at: <http://www.bmj. com/ content/315/7121/1533> (Accessed 18 April 2014).

Elgendy, K. (2012). Sustainable development and the built environment in the Middle East: challenges and opportunities. *Middle East Institute.* [online] Available at: <http://www.mei.edu/content/sustainable-development-and-built-environment-middle-east-challenges-and-opportunities> (Accessed 28 April 2014).

Elliot & Associates. (2005). Guidelines for conducting a focus group. [online] Available at: <http://assessm ent.aas.duke.edu/documents/ How _to_Conduct_a_Focus_Group.pdf> (Accessed 30 April 2014).

Eurostat European Commission. (2009). Sustainable development in the European Union: 2009 monitoring report of the EU sustainable development strategy. *Eurostat European Commission.* [online] Available at: <http://epp.eurostat.ec.europa.eu/ cache/ITY_OFF PUB/KS-78-09-865/EN/KS-78-09-865-EN.PDF> (Accessed 30 April 2014).

Evelyn, M.I. & T. Tyav. (2012). Environmental pollution in Nigeria: the need for awareness creation for sustainable development. *Journal of Research in Forestry, Wildlife, and Environment*, 4 (2). [online] Available at: <http://www.ajol.info/index.php/jrfwe/ article/view/84726> (Accessed 27 April 2014).

Forstater, M. (2003). Public employment and environmental sustainability. *Journal of Post Keynesian Economics*, Spring, 25 (3), 385-406.

Furusawa, K. (2012). Towards a sustainable civilization and society: socio-cultural perspective from Japan. *Encyclopedia of Life Support Systems.* [online] Available at: <http://www.eolss.net/sample-chapters/c16/e1-57-51-00.pdf > (Accessed 20 April 2014).

Garcia, L. E. (2014). Sustainable development and the private sector: a financial institution perspective. [online] Available at: <http://environmen t.research.yale.edu/ documents/ downloads/0-9/101garcia. pdf> (Accessed 18 April 2014).

Government of Ireland. (2014). Our sustainable future. [online] Available at: <http://www. environ.ie/en/Environment/SustainableDevelopment/PublicationsDocuments/FileD own Load,30454,enpdf> (Accessed 30 April 2014).

Greenland, S. & K. O'Rourke. (2008). Meta-Analysis." In *Modern Epidemiology*, 3rd ed. Edited by Rothman, KJ, Greenland, S., Lash, T.., Lippincott Williams & Wilkins, 2008.

Guardian News and Media Limited. (2012). The role of the private sector in sustainable development. *Guardian News and Media Limited.* [online] Available at: <http://www. theguardian.com/sustainable-business/un-sustainable-development-private-sector. 2012> (Accessed 5 May 2014).

Guijt, I. & A. Moiseev. (2001). Resource kit for sustainability assessment. *IUCN.* United Kingdom: Gland and Cambridge.

Gulick, R.V. (2004). Consciousness. *Stanford Encyclopedia of Philosophy*. [online] Available at: <http://plato. stanford.edu/entries/consciousness/> (Accessed 3 May 2014).

International Institute for Sustainable Development. (2013). National strategies for sustainable development challenges, approaches and innovations in strategic and co-ordinated action, a 19-country study.

International Institute for Sustainable Development. [online] Available at: <http://www.iisd .org/ measure/gov/sd_strategies/national.asp> (Accessed 6 May 2014).

International Telecommunication Union. (2012). Smart sustainable development model. [online] Available at: < http://www.itu.int/ITU-D/emergencytelecoms/ initiatives/ SSDM.pdf> (Accessed 2 May 2014).

Johansson, B., A. Dagman, E. Rex, T. Nyström, M. Wedel, J. Stahre, & R. Söderberg. (2012). Sustainable production research: awareness measures and development. *OIDA International Journal of Sustainable Development,* 04(11), 95-104. [online] Available at: <http://papers.ssrn. com/sol3/papers. cfm?abstract_id=2145654> (Accessed 3 May 2014).

Kingdom of Bahrain E-Government Portal. (2014). Protection of environment and natural resources. *Kingdom of Bahrain E-Government Portal.* [online] Available at: <https://www. bahrain.bh/wps/ portal/!ut/> (Accessed 5 May 2014).

Legrand Group. (2014). Sustainable development: definition, background, issues, and objectives. *Legrand Group*. [online] Available at: <http://www.legrand.com/ EN/sustainable-development-description_12847.html> (Accessed 6 May 2014).

Leiserowitz, A., R. Kates, &T. Parris. (2005). Do global attitudes and behaviors support sustainable development? *Environment,* 47(9), 22-38.

Mathey, J. (2014). Five elements of sustainability. [online] Available at: <http://www.matthey. com/ sustainability/why_ sustainability/ five_ key_elements > (Accessed 1 May 2014).

McKeown, R. (2002). Challenges and barriers to ESD. *ESD Toolkit*. [online] Available at: <http://www. esdtoolkit.org/ discussion/ challenges.htm 2002> (Accessed 21 April 2014).

Michalos, A., H. Creech, C. McDonald, & P.M.H. Kahlke.(2011). Knowledge, attitudes and behaviours concerning education for sustainable development: two exploratory studies. *Social Indicators Research,* 100, 391-413. [online] Available at: <http://www.iisd.org/ publications/measuring-knowledge-attitudes-and-behaviours-towards-sustainable-develo pment-two> (Accessed 25 April 2014).

Organisation for Economic Cooperation and Development. (2008). Gender and sustainable development: maximising the economic, social, and environmental role of women. *Organisation for Economic Cooperation and Development (OECD)*. [online] Available at: <http://www.oecd.org/social/40881538. pdf> (Accessed 25 April 2014).

Owolabi, H.O., E.F. Gyimah, & M.O. Amponsah. (2012). Assessment of junior high school students' awareness of climate change and sustainable development in Central Region, Ghana. *Educational Research Journal,* 2(9), 308-317. [online] Available at: <http://www. resjournals. com/ ERJ> (Accessed 25 April 2014).

Public Works and Government Services of Canada. (2010). Planning for a sustainable future: A federal sustainable development strategy for Canada. *Her Majesty the Queen in Right of Canada, represented by the Minister of the Environment.* United Kingdom.

Rafique, M. (2014). Govt's sustainable development policies win praise. *24/7 News.* [online] Available at: <http://www.twentyfourse vennews.com/bahrain-news/govts-sustainable-development-policies-win-praise> (Accessed 28 April 2014).

Razman, M.R., J.M.. Jahi, S.Z.S. Zakaria, A.S. Hadi, K. Arifin, K. Aiyub & A. Awang. (2012). Environmental awareness towards sustainable development through the principle of transboundary liability: international environmental law perspectives. *Research Journal of Applied Sciences,* 7, 194-198.

Research and Policy in Development. (2009). Research tools: focus group discussion. [online] Available at: <http://www.odi.org.uk/publications/5695-focus-group-discussion> (Accessed 30 April 2014).

Shastri, R. (2005). Environmental awareness and sustainable development. [online] Available at: <http:// ideas.repec.org/p/wpa/wuwpo t/0504001.html#cites> (Accessed 25 April 2014).

Sillitoe, P. (2014). Sustainable development: an appraisal of the Gulf Region. [online] Available at: <http:// www.bokus.com/ bok/9781782 383710/sustainable-development/> (Accessed 24 April 2014).

Smith, C. (2009). Sustainability and water resources in the Middle East. *E-International Relations.* [online] Available at: <http://www.e-ir.info/2009/11/09/sustainability-and-water-resources-in-the-middle-east/> (Accessed 22 April 2014).

Social Media Week. (2014). Open yet unconnected? social media for sustainable development. *Social Media Week.* [online] Available at: <http://socialmediaweek. org/lagos/ events/ ?id=54449> (Accessed 23 April 2014).

Stenberg, J. (2001). *Bridging Gaps – Sustainable Development and Local Democracy Processes.* Gothenburg.

Stratos Inc. (2004). United Kingdom case study analysis of national strategies for sustainable development. *Stratos Inc.* [online] Available at: <http://www.iisd.org/pdf/2004/ measure_sdsip_uk.pdf> (Accessed 22 April 2014).

Svensson, E. (2012). Achieving sustainable lifestyles? sociocultural sispositions, collective action and material culture as problems and possibilities. [online] Available at: <http://www.researchgate.net/pub lication/241731533_Achieving_ sustainable_ lifest yles_ Sociocultural_dispositions_collective_action_and_material_c > (Accessed 25 April 2014).

Syracuse Onondaga County New York. (2011). Sustainable development plan. [online] Available at: <http://future.ongov.net/?page_id=22> (Accessed 25 April 2014).

Ugurlu, N.B. & E. Aladag. (2004). Natural resources and education for sustainable development. [online] Available at: <http://www.herodot.net/conferences/ayvalik/papers/environ-04.pdf> (Accessed 25 April 2014).

United Nations Department of Economic and Social Affairs. (2013). World economic and social survey 2013: sustainable development challenges. *United Nations Department of Economic and Social Affairs.* [online] Available at: <http://sustainable development.un.org /content/ documents/ 2843W ESS2 013.pdf> (Accessed 26 April 2014).

United Nations Development Program in Bahrain. (2012). Inclusive sustainable development in Bahrain. [online] Available at: <http://www.bh.undp.org/content/bahrain/ en/home/ ourwork/inclusive-sustainable-development/overview.html> (Accessed 20 April 2014).

United Nations Environment Program. (2014). Children and youth in sustainable development. *United Nations Environment Program.* [online] Available at: <http://www.unep.org/ Documents.multilingual/Default.asp?DocumentID=52&ArticleID=73&l=en> (Accessed 21 April 2014).

United Nations Scientific and Cultural Organization. (2014). Education for sustainable development: United Nations Decade 2005-2014. [online] Available at: <www.unece.org/env/ esd/events.../UNESCO.JAN06.doc. 2005> (Accessed 21 April 2014).

Virtanen, P.K. & S. Saarinen. (2012). How to integrate socio-cultural dimensions into sustainable development: Amazonian case studies. *Int. J. Sustainable Society,* 4 (3), 226-239.

World Bank. (2013). Doing business in Bahrain. *World Bank.* http://www. [online] Available at: <Doing business.org/data/exploreeconomies/bahrain/> (Accessed 21 April 2014).

World Commission on Environment and Development. (2008). Our common future: report of the World Commission on Environment and Development. *World Commission on Environment and Development.* http://www. [online] Available at: <http://www.un-documents.net/ocf-02.htm#III.2> (Accessed 21 April 2014).

8

The Growing Thirst of the United Arab Emirates: Water Security Stresses that Challenge Development

Rachael McDonnell

1. Introduction

The United Arab Emirates (UAE) is rapidly recovering from the economic downturn of 2008 with many new developments, particularly in Abu Dhabi and Dubai Emirates, highlighting its remarkable growth is back on track. The rising GDP from many economic sources not just hydrocarbons confirms its position as an increasingly important global player in the worlds of trade, transport and tourism. The latest announcements for urban development plans for Dubai and Abu Dhabi reveal thousands more homes, amenities such as golf courses, new business hubs, even larger shopping malls, expansion of transport infrastructure and many more tourist destinations. Whilst these new projects will bring economic diversification, growth and employment, the concomitant increasing population will put severe strains on existing systems providing water, energy and food security. The natural water provision in the arid climes of the Gulf states ensures water is one of the fundamental limiters of economic development yet with the provision of new resources that are manufactured – desalination and treated wastewater – this is being overcome. Today water is provided 24 hours a day, 7 days a week so would suggest that, using the definition of Grey et al (2013) "*water security is a tolerable level of water-related risk to society*', the current risk is relatively low and the UAE is reasonably water secure. Yet the underlying state of the water resource balance equation and long-term dynamics in available supplies and demands paint a quite different picture. Greatly improved sustainable management will be required in addition to technology options with associated large capital and operational expenditure, to maintain this water security state.

In the UAE the governance and management of water systems today are very different from those traditional community-based arrangements of the era before oil was discovered (Wilkinson, 1997; Lightfoot, 2000). They were gradually replaced as the emirates developed economically especially with new hydrocarbon revenues supporting the drilling of new and deeper wells. The founding of the federal state in 1971 brought with it a constitution that established formally that natural resources, and so water, are the public property of each emirate. The subsequent developments within each Emirate, each following their own approaches to supplies, management and regulation, have brought increasingly complex water provisions.

This chapter will explore the water resource systems in the UAE today, its management and governance frameworks and control the various components of supply and demand. Understanding the institutional, legal and economic dimensions of water is often overlooked, yet is an important aspect of the water security debate especially in a federal state such as the UAE. The chapter will also explore the growing dependence on 'manufactured' water over the last two decades, which bring new economic, energy and water challenges. These new sources of water have completely changed the dynamics and magnitude of the country's water resource systems. They have supported and influenced growing demand patterns that far outstretch the available natural groundwater supplies. They have also changed society's relationship with water leading to landscapes and patterns of water use that little reflect the desert world or traditional practices (McDonnell, 2013). The forecasted gap in water availability from current supply systems against demand ensure that managing this resource wisely is going to be critical and strategic if the developments proposed are sustainable over the long term. The waterscape in the UAE and all the Gulf states challenges our perceptions of water security and the development possible in any country where conditions are naturally arid.

2. The Water Resource Balance Today

The starting point for understanding a country's water security is a review of its resource base and the nature of demands across the various sectors. These influence when and where water is available and required.

2.1. Sources of Water

Groundwater has traditionally been the main source of water for all uses in the UAE, and its location and availability has greatly influenced settlement and farming systems in the past. The groundwater systems of UAE are predominantly based on recharge events 9,000-6,000 years Before Present (BP) and 32,000-26,000 years BP with minimal natural replenishment today even with attempts to increase aquifer recharge through the building of dams (Woods and Imes 1995; Al-Sharhan et al., 2001; Imes and Clark 2006; Rizk and Al Sharhan, 2008). This means that groundwater abstractions, which account for 51% of total water supply, are essentially mining (Ministry of Water and Environment MOEW, 2010). Of the total groundwater available, only 3% of it is fresh with varying degrees of salinity affecting most of the resources. In recent years human activities have added to the water quality issues, with ground-water contamination arising from agriculture's extensive use of pesticides and fertilizers and over-abstraction leading to even greater salinity in coastal and inland aquifers.

In the past groundwater was managed through well and *aflaj* (groundwater-fed irrigation channel) resource systems in oases and coastal villages, with local management and allocation rules developed and enforced to ensure both sustainable quantity levels of consumption and pollution prevention (Lightfoot, 2000; Wilkinson, 1977). The community level governance system based on owner-participatory approaches and common customs and value systems met many of the criteria for "good water governance" defined today. Aflaj followed a market-based approach in which residents gained rights to the water produced by the irrigation channel infrastructure development, following their investments in money or labour leading to allocation of shares in the water resource. These water rights were tradable with auctions supporting maintenance of the system.

The fast economic development of the last thirty years with a multitude of new sectoral demands as well as an ever increasing and affluent population has lead to aquifers being over-exploited and the rapid development of alternative fresh water sources. Today in addition to groundwater, the other main water sources used in the UAE are:

- desalinated water accounts for 37% of total water supply and is used mainly for potable water use, in addition, small quantities of fresh and desalinated groundwater are used for domestic and industrial purposes particularly in Northern Emirates, and
- reclaimed water supply which accounts for 12% of water supply and is used mainly for amenity and landscaping within the urban areas (MOEW, 2010).

Desalinated supplies, from 70 seawater and inland water plants using thermal and membrane technologies, is the primary source of potable water. The desalination plants are predominantly co-generation with electricity produced from turbines driver by steam which this is then captured and cooled to form water. The energy consumption is variable ranging from 15.7-5kWh depending on the technology used, with membrane technology about 85% less than thermal systems (MOEW, 2012). Whilst in some parts of the world today this membrane technology, involving filtering of the seawater under high pressure, are most used, thermal systems in the Gulf still dominate. This results from the energy efficiencies of co-generation production as well as their proven operational robustness under the exacting conditions of the Gulf summer.

Desalinated water is produced and distribute by four largely Emirate-level organizations with capacity varying greatly across the Emirates (see Figure 1) .Over the last few years, the development and expansion of Abu Dhabi's water production and distribution systems has also helped to meet the needs for both water (and electricity) in neighbouring emirates, Ajman, Umm al Quwain, Ras al Khaimah and Sharjah. These Northern emirates have markedly lower GDPs as they lack the hydrocarbon wealth of Abu Dhabi[1] and their investments have been limited in major and expensive infrastructure projects despite their increasing water demand. The transfer of over 56 million cubic metres in 2012 (Abu Dhabi Water and Electricity

Company, 2012) from Abu Dhabi's water production systems has provided a new, important and reliable supply of water to support their economic development, continuing the long tradition of extending Abu Dhabi's 'fatherly care' to other emirates.

Figure 1 **Desalination Water Capacity and Percentage per Emirate**

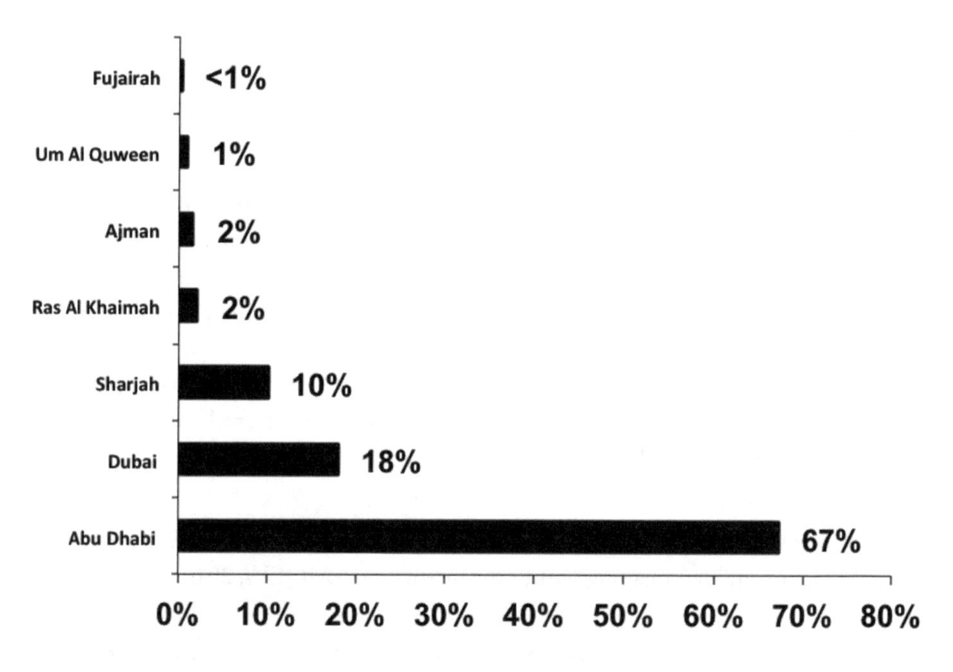

Source: MOEW, 2010

The third source of water, reclaimed water (otherwise known as treated sewage effluent) is increasing in availability as a consequence of the rise in the population of the UAE. There is variable availability of reclaimed water depending on the presence or absence of sewerage networks and wastewater treatment plants. In the major urban centers of Dubai and Abu Dhabi, sewage water is collected through networks or tankers, and treated at a range of size of plants. In many of the new housing mega-projects sewage is collected and treated locally and then used within the developments in landscaping. Whilst there are limits to the use of this water because of its potential harm on people or the environment, there are advantages in its use as a substitute for better quality water, particularly expensive desalinated water and fresh groundwater. It is used predominantly for landscaping but with increasing clarification and development of regulations, the possibilities of its use in other areas are gradually being realised. As the distribution infrastructure and regulatory systems are developed the main limiter to extensive use is likely to be public perceptions.

Turning to the demand side of the water balance, it arguably this that is the real and seemingly intractable challenge to sustainable development. The UAE has one of the highest per capita rates in the world, being driven up over the last fifty years by agricultural expansion, population growth, high standards of living and industrialization. The current per capita urban water consumption (domestic, industrial and commercial, amenity and

2.2. The Nature of Water Demand

Figure 2 **UAE Sectoral Distribution of Water Use in 2008**

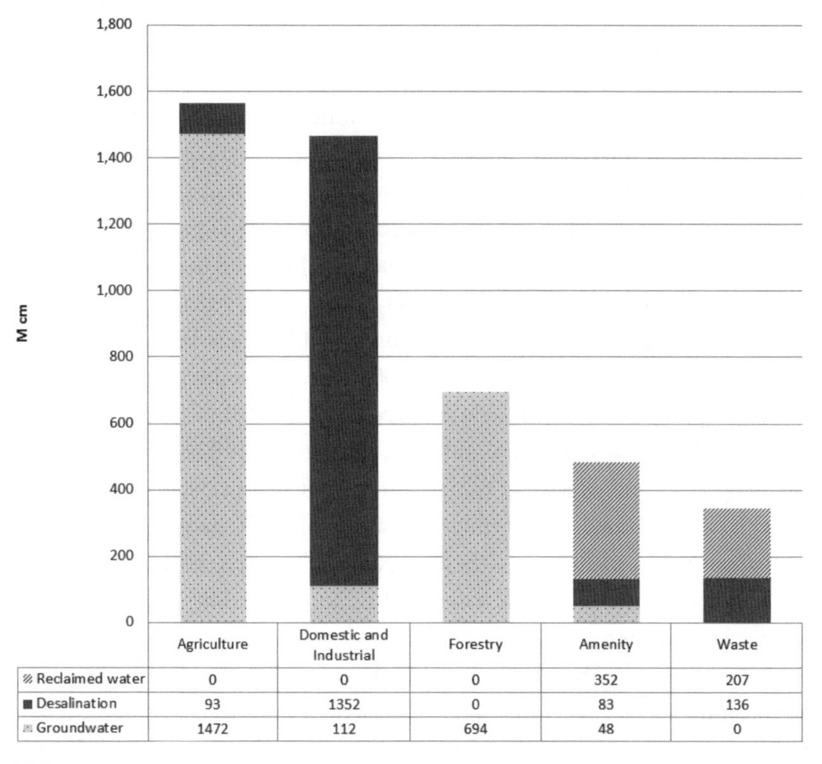

	Agriculture	Domestic and Industrial	Forestry	Amenity	Waste
Reclaimed water	0	0	0	352	207
Desalination	93	1352	0	83	136
Groundwater	1472	112	694	48	0

Source: MOEW, 2010

governmental uses) is about 753 litres per capita per day, with household consumption accounting for about half (364lpcd) of all desalinated water use (MOEW, 2010). With population growth from 178,600 in 1968 to over 6 million in 2012 and industry and manufacturing (excluding oil) going from almost zero in 1965 to 22% of GDP in 2008, the demand for water is ever increasing. However, even given these rapid urban and economic developments, agriculture is still the single largest user of water and almost 60% of all UAE's water (principally groundwater) is used for agriculture, forestry and urban greenery (MOEW, 2010).

The use of extensive water supplies in irrigation is visible to any visitor to the country with lush urban areas, tree lined roads stretching out into the desert, vibrant green small-holder farms and huge date palm plantations. As national prosperity has increased the Federal and several Emirate governments have encouraged afforestation and agriculture through a generous program of subsidies. The "greening program" which included parks, roadside afforestation and lawns, shelter belts and rural forests as part of game reserves, now consumes large volumes of groundwater or treated wastewater. In most homes and some farms, the water used is desalinated which means the greening is at a huge energy and economic cost that is largely borne by the government.

2.3. The Special Case of Agriculture

For water managers in the UAE, the irrigation of many thousands of hectares of land is a major challenge especially given the irreversible abstraction from aquifers that it brings. Agricultural policies and production are driven by both cultural values and the need to develop a base level of local food security. Traditionally agriculture was limited to vegetable and date groves in oasis areas fed by springs, or the rearing of camels, sheep and goats on rangeland. With changes in the economy following the discovery of oil the rapid expansion of agriculture was facilitated through many different subsidies and unlimited and free access to groundwater reserves. As a result the farmed area increased from 3,000 hectares (ha) in 1968 to over 100,000 ha by 2008 with a particular expansion in tree crops (mostly date palms), vegetables and forage. At the same time livestock farming also grew with numbers of livestock increasing from about 0.5 million to 3.3 million between the 1980 and 2008. The concomitant increase in forage crop production to support the livestock industry accounted for 39% of farmed area in 2008 and accounted for 45% of total agricultural water use (Pitman et al, 2009; MOEW, 2010).

The period of rapid expansion came to an abrupt halt in 2000 and thereafter slowly declined with the cultivated area shrinking at a rate of 4,200 ha per year. This is likely to have come about through changes in the Federal and Emirate-level policies on agricultural subsidies, shortages of good groundwater quality, and increasing production costs; all of which reduced profitability.

The actual amount of water used in agriculture is influenced by a number of different variables including crops grown, management practices and the irrigation technology used. The latter has been an important focus for water conservation efforts since the mid-1990s with more than 97% of cultivated areas equipped with modern irrigation technology that included sprinkler, drip and bubbler water application systems by 2008. This has lead to a reduction in more than 30% of water per hectare less over the last decade but further developments are needed in irrigation management and crop preferences if the impact of agriculture is to be reduced to more sustainable levels (Pitman et al, 2009; MOEW, 2010).

3. Water Governance, Laws and Management

The sustainable and effective management of a scarce resource such as water requires competent governance, legal and regulatory frameworks systems to be in place. In the UAE as a result of the 1971 constitution which established a federal state, institutions, legislation and regulations governing natural water resources are engaged mostly at the Emirate level.

The federal legislature and executive retain authority for national policy/strategy overview. There are thus different organizations in the seven Emirates with direct responsibilities for managing the water resources whilst various ministries are primarily responsible for strategic level decision-making in particular parts of the water system.

3.1. Water Supply Governance

The three main sources of water in the UAE are quite different in nature and origin, it is therefore unsurprising that each has differing institutional arrangements and counterparts across the seven emirates. With groundwater the federal level MOEW is responsible for protection of the resource but day to day operations and abstractions is managed through organizations such as Environment Agency Abu Dhabi and departments within the municipalities in the other emirates.

For desalinated water, federal authority is under the Ministry of Energy, with different specialist organizations involved at the Emirate level in the production, transmission, distribution and control of this. With the exception of Abu Dhabi these authorities are government owned although the private sector is involved to various degrees in design, construction and operation. In Abu Dhabi a system of highly specialized organizations has been established and the private sector is involved to a much greater extent here through public-private-partnerships, particularly in the area of desalination water production. In addition to these large Emirate based authorities, there are an increasing number of smaller organizations, often associated with large developments, which are licensed to produce and desalinate their own water. Some, such as Dubai's Palm Water, are licensed for extended periods, whilst others such as the Tourism Development and Investment Company in Abu Dhabi have a fixed period of operation.

With reclaimed water a range of institutional provisions have evolved over the last few decades and these are predominantly government agencies. At the Federal level jurisdiction and services were transferred to the MOEW in 2009 and at the moment this is not an active area of legislative or regulatory development with responsibilities still to be finalized. At the emirate level the governance is variable. In some of the emirates, one principal organization is involved in collecting treating and disposing of wastewater, whilst in others there is a mix of various agencies and departments responsible for these different areas. In recent years the private sector has become increasingly involved in managing sewage services through public/private partnerships with the first concession granted in Ajman through the establishment of Ajman Sewerage Private Company Limited. In addition to the principal Emirates-level organizations there are an increasing number of smaller private companies involved in the reclaimed water sector. Recent moves by mega-developers and other organizations to manage their own decentralized provisions have meant further private sector involvement. Local companies such as Nakheel, Emaar, Dubai Holding and Sports City are constructing the wastewater infrastructure for their projects, including sewage treatment plants, sewerage network, pumping stations and reclaimed water distribution networks to serve these communities. In addition in some municipalities the application and use-management of reclaimed water is outsourced to various private sector companies.

As the preceding discussion has highlighted, the institutional framework for water in the UAE is relatively fragmented with many different authorities/agencies involved. Whilst managing water at the Emirate level has not been without its own success, the present setup has led to overlapping roles, whilst gaps in key areas of strategic or operational planning fall between the various water sources and/or water uses. This is not an ideal institutional framework for managing an increasingly water stressed situation with rapidly developing economies dependent on secure supplies of water. Continuing to manage water risks will require greater strategic coordination across and between federal and Emirate levels.

3.2. Water Laws and Regulation

In addition to the institutional setup of a country, laws, regulations, standards and their enforcement are important in ensuring the protection of human and environmental health as well as economic efficiency. They give direction, transparency and clarity in many areas such as institutional responsibilities and roles, legal clarification on property rights, and standards for a particular environment or sector. In the UAE the 1971 constitution divides legal rights and responsibilities between the Federal and the Emirate levels clearly. Whilst water is not mentioned explicitly, by implication of some of its provisions (Articles 23, 120, 121, 122) natural (water) resources are the public property of the Emirates. This provides a solid legal backdrop to the current system of the Emirates' exercise of regulatory authority over water resources abstraction, use and protection from pollution. This is further confirmed by the laws in effect in the different Emirates such as Abu Dhabi Law No. (6) 2006 on Regulation of Drilling of Water Wells. As a result, legislation and regulations governing natural water resources are engaged mostly at the Emirate level, with the Federal legislature and executive retaining authority for national policy/strategy overview. This is particularly in the area of environmental protection, such as through Federal Law (24) 1999, with responsibility given to MOEW.

Most of the emirate level groundwater laws in place attempt to control the drilling of new wells, or the further development of existing ones through requirements for licensing. Some laws include details on the actual control measures (metering, pumps etc), and on the data to be provided by well-owners such as the use of the water and estimated rate of abstraction. The most difficult area for enforcing the water laws is in regulating existing wells. These wells, and traditional groundwater rights which pre-date regulatory legislation, play a significant role in much of the country's rural areas. Their use is grounded on the deep and diffuse conviction of the owners that, regardless of the articles of the Constitution, groundwater is the property of the landowner, for him to extract and dispose of as he sees fit. As a result, managing their rights (and convictions) in the context of a new law introducing regulation of water resources abstraction and use, raises delicate "expropriation" and so sustainable development issues.

In contrast to groundwater, the legal status of desalinated and reclaimed water is not explicitly defined in the constitution. Whilst subsequent laws have been enacted establishing the organizations and regulatory structures responsible for these two areas of water resource development, the supplies themselves are in effect the legal property of the relevant producer, and so are for it to dispose of and allocate as seen fit. For both reclaimed and desalinated water there is a need through regulations to protect consumers in terms of both health and tariff structures, as well as ensure little environmental damage. The main organizations have developed various water regulatory provisions for controlling brine discharges from desalination plants, the qualities of reclaimed water and their permitted uses, as well as the minimum/maximum limits for a range of chemical element values in potable water. The regulations vary widely between the emirates and the effectiveness of implementation and enforcement also differs across the authorities.

As with water governance, any review of the legal and regulatory provisions for water in the UAE highlight the need for harmonization both across the various water supply systems and between the emirates to protect the resources both in qualitative and quantitative terms. Without further development of clarity and coordination including national level standards and regulations future resource availability and sustainable development will be challenged.

3.3. Water Resources Management

Leading on from the governance and laws and regulations, a fundamental component in balancing the competing demands and resource base is the planning and operations involved in water resources management. Reflecting the governance and legal systems in the UAE this is undertaken predominantly at the Emirate level with many different organizations involved.

To date the main emphasis of management has been on developing new supplies. Innovative technologies to produce more water than the natural resource base have been central to these efforts. Desalination, with its complex production and transmission systems, has become the main stay for drinking water but storage is limited ensuring only 2-5 days of supplies are available at any time. This means that in the case of an emergency, such as an oil spill or terrorism incident in the Gulf affecting production, groundwater is the only strategic reserve to serve the direct needs of the population. For reclaimed water there have been large investments into new treatment plans and distribution networks to further develop this area of supply. The hot and humid conditions as well as presence of saline water intrusion in some parts of the sewerage network can both help and hinder the treatment processes across the three stages.

The development of both desalination and reclaimed water capabilities has required enormous investments. These have been raised from government funds, through the capital markets including bonds or from private sector investment with capital expenditure exceeding many billions of dollars. For operational costs, energy inputs are an important component with gas from neighbouring Qatar being particularly important (McDonnell, 2013). The total estimated annual expenditure in desalination and reclaimed water production over the last five years is highlighted in Table 1.

Table 1 Estimated Annual Capital and Operational Expenditure (US$m)

Project	2009	2010	2011	2012	2013
Capital Expenditure					
Desalination	1870	1809	1196	1089	1470
Water distribution network	101	117	133	148	162
Wastewater network	502	558	620	602	669
Wastewater treatment	361	415	478	481	554
Operational Expenditure					
Water	1175	1361	1474	1581	1687
Wastewater	101	148	202	262	322
Total Expenditure	4110	4408	4103	4163	4864
Cumulative Expenditure		4408	8511	12674	17538

Source: MOEW, 2010

In terms of managing the demand of water, the main policy instruments used are awareness raising campaigns to inspire reduced use. These have been particularly aimed at the use of water in gardens as the example shows in Figure 3. To date the water provided to homes and hotels is desalinated so the large irrigation of gardens and lawns and the filling of swimming pools is based on some of the most expensive and highly subsidized water possible.

Figure 3 Poster Campaign to Reduce Water Use

Source: http://water.heroesoftheuae.ae/en (accessed 10/3/2014)

To a lesser extent economic pricing has been used with around only 22% of the total water used in the UAE each year paid by customers. Groundwater and reclaimed water that account for 2.9 km^3 or 63% of all water supplies are provided free with tariffs only on desalinated water, and even then these are highly subsidized to nationals particularly in Abu Dhabi Emirate. Water tariffs for non-nationals range from UAE Dirhams 2.2 (US$ 0.6) per m^3 in Abu Dhabi to as much as AED 8.8 per m^3 in the others. In addition commercial and industrial consumers who account for 0.7 per m^3 or 40% of desalinated water consumption pay a similar tariff to non-nationals. The cost of producing desalinated water, however, is much higher than the tariffs charged and because many consumers pay less than the full cost, the implicit subsidy is about AED 6 (USD 1.64) per m^3. Annually this amounts to about AED 9.5 billion (USD 2.59 billion) (MOEW, 2010). For reclaimed water although it is provided free of charge, the average costs are AED 4.8 (USD 1.31) per m^3 to produce. Consequently the implicit subsidy is about AED 4.2 billion (USD 1.14 billion) for the 0.86 km^3 produced each year (MOEW, 2010). The result is a large financial commitment by the various emirate level governments to ensure water security for its citizens and industries.

4. Water Security Sustainable Development Challenges

This chapter has focused on the special case of water, a vital resource in all Gulf States, with its development and use tightly linked to energy and food production. The three sources of water each bring possibilities for supporting both current and future developments in the UAE however a series of challenge that, to date, have not been addressed are likely to affect the ability to meet future demands. The case has already been made for more coherent water governance, laws and management to bring efficiencies and effectiveness to planning, operations and developments. Other challenging areas however also need to be explored.

From the discussions so far, it is clear that for the UAE the strategic water reserve is groundwater. In the case of an emergency such as failure in desalinization facilities, groundwater is the only reserves that can be used for drinking by humans. Yet fresh and brackish groundwater is being used at an alarming rate with little protection to reserves being afforded through laws and regulations. The agricultural lobby is particularly strong politically even though much of the crop and livestock production systems, and so water use, are based around small units of 'hobby' or weekend farming. Their contribution to national food security is limited yet their irrigation practices challenge the basic resource water security is based on. There is a need to increase the protection of groundwater systems and new zoning measures being reviewed in Abu Dhabi offer possibilities of effecting this. Other ideas under discussions by the Environment Agency Abu Dhabi include linking water use licensed to the crops/areas under production for individual farms (Environment Agency Abu Dhabi, 2014). Perhaps the most effective means of protecting groundwater systems is to rationalize even further the subsidies paid to farmers for crop and livestock production. Replacing these with social support for rural livelihoods and using food futures to help reduce the risk of global food market impacts may not only be more cost-effective but also increase future water security at the same time.

For desalinated water the challenges to future production development are centred on environmental and economic areas. The environmental impact of increasing numbers of desalination plants of Arab Gulf State countries is predominantly directly linked to the levels of hot salty brine returning to the sea. The average salinity of the Gulf these days is around 45,000 parts per million but in the areas of outflows from desalination plants these levels can rise rapidly. Often forgotten is also the impact of heat of brine, that whilst it has been cooled significantly from the temperature of steam, it is still often tens of degrees different to the conditions of the receiving waters. In both cases the effluent discharge effects can impact severely local ecosystems, particularly affecting fish stocks and corals. With increasing numbers of desalination plants being built or planned by all the Gulf states the ability of the marine environment to absorb the cumulative impacts is likely to be stressed, compounding the impacts of coastal developments taking place.

The second challenge is centred on the financing of desalination developments. As Table 1 highlighted, the capital and operational costs are immense and these are little offset by income from consumer tariffs. Given that citizens and residents of the UAE do not pay tax, the funds for desalinated water development either directly or indirectly come from the economic activities of the various emirate governments. For Abu Dhabi and so some of the Northern Emirates, this is made possible by income from oil exports which is affected by global dynamics in markets. Uncertainties in oil prices from changes in production and demand such as the shale and fracking developments in the USA, opening up of Iran and Iraq export supplies, or the economic growth or decline in key markets in Europe and Asia directly affect income and government budgets. Unlimited capital investments and consumption subsidies are impossible meaning both water supply system development and demand need to be managed so that the financial inputs are more sustainable.

Arguably the biggest challenge results from the rapid change in the relationship between society and nature in the last few decades. Traditional community water-based systems that were overseen by local/tribal leaders have been replaced by emirate level governments supplying directly water to homes, farms and businesses. The previous local rules, governance systems and risk management methods that ensured a community's survival have been replaced by invisible production systems and infrastructure. Urban citizens in particular now experience water through bottled provisions, non-stop tap flows and irrigated surroundings. With little paid financially for the water there has also been a loss in its perceived value by society hence the extravagant water use practices of so many. This change in mindset and relationship with water is echoed across many different dimensions and levels within UAE societies. For example posters advertising new residential and commercial developments levels show a lush green urban and both expatriates and citizens expect this today as part of their local landscape. The frugal water management practices of people two generations ago have long been disused and the risks and responsibilities of individuals and communities to manage the natural water scarcity have been passed to governments.

Changing this relationship to one that supports long term water security and sustainable development seems an intractable problem at present. The usual information, economic and policy instruments seem blunt tools to bring about this much needed change in a society that is increasingly metropolitan and sophisticated, and overwhelmingly international. In many ways this new relationship between water and society in an arid state that relies on manufactured water, challenges us to think outside normal frames of reference. Should we think about water as solely a produced commodity or does it have unique set of properties that need to be acknowledged and which are central to a society's existence? The changes in the UAE in the coming years as new developments are built and populations expand, will need to be examined in more detail to help understand how water and society, and so political and economic relationships, are changed with the need to provide and consume water.

5. Conclusions

Water management is a fundamental component of sustainable development in the Arab Gulf States. The natural arid conditions have ensured water security has always been part of the lives of communities and citizens. In recent decades new water sources have brought new possibilities for resource management and users. At the same time increasingly complex governance and legal/regulatory frameworks have been put in place as governments replaced community level management. With this has brought a shift in responsibility for all aspects of water production and distribution in terms of quality, quantity and cost of the supplies. Sustainable development in the future will increasingly rely on individuals taking back some of that responsibility to ensure their demands are not pushing supply systems and the affected economic and natural environments to limits that are harmful.

Note

1 In the UAE, natural resources are the property of each emirate, rather than the country.

References

Abu Dhabi Water and Electricity Company 2012. *ADWEC statistical leaflet 2012*, http://www.adwec.ae/Statistical (Accessed 3.4. 2012)

Alsharhan, A.S., Rizk, Z.S., Nairn, A.E.M., Bakhit,D.W., and Alhajari, S.A., 2001. *Hydrogeology of an arid region: The Arabian Gulf and adjoining areas*. Elsevier Publishing Company, Amsterdam.

Grey, D., Garrick, D., Blackmore, D., Kelman, J., Muller, M. and Sadoff, C. 2012, Water security in one blue planet: twenty-first century policy challenges for science, doi:10.1098/rsta.2012.*0406 Phil. Trans. R. Soc.* A, 13 November 2013 vol. 371 no. 2002 20120406

Imes, J.L, and Clark, D.W. 2006. *National Drilling Company-U.S. Geological Survey Technical Services Administration Report*, 2006-001,001.

Lightfoot, D.R. 2000. The origin and diffusion of qanats in Arabia: new evidence from the northern and southern Peninsula. *The Geographical Journal*, 166 (3): 215-226.

Pitman, K., McDonnell, R. and Dawoud, M. 2009. *Abu Dhabi Water Resources Master Plan.* Environment Agency Abu Dhabi, Abu Dhabi.

Rizk, Z.S., and Alsharhan, A.S., 2008. *Water resources in the United Arab Emirates.* Ithraa Publishing and Distribution, Sharjah, United Arab Emirates (in Arabic).

Wilkinson, J. C. 1977. *Water and tribal settlement in south-east Arabia: study of the aflaj of Oman.* Oxford: Clarendon Press.

Woods, W. and Imes, J. 1995.How wet is wet? Precipitation constraints on late Quaternary climate in the southern Arabian Peninsula. *Journal of Hydrology*, 164, (1–4), 263–268

9

Food-Water and Food Supply Chains: A Cornerstone of Sustainable Development in the Gulf Cooperation Council

John Anthony Allan, Mark Mulligan and Martin Keulertz

1. Introduction

The purpose of the chapter is to highlight the importance of food supply chains in understanding Gulf Cooperation Council (GCC) water security. Food supply chains are the metabolism of global food trade. In hyper-arid regions such as the GCC, food supply chains provide societies with imported food to keep the economies food-secure. However, food supply chains do not only trade food from one point of the world to another but also carry water. 90 per cent of the water needed by an individual or a national economy is embedded in their food consumption. This water will be called food-water in this analysis. Food requires transpired water to produce it and water is the main limiting factor in the GCC region to increase food production. The GCC economies are thus among the leading food importing economies in the world. Their reliance on food supply chains is therefore evident.

The aims of this chapter are twofold. It will first introduce the food supply chain framework to the reader to lift the topic of water resource management from a mere issue of national strategy into the arena of international trade. The authors of this chapter view farmers and thus the private sector as pivotal to global and regional food and water security and to installing measures to steward water ecosystem services. Given the increasingly globalised political economy of agriculture, the food supply chain framework will enable the reader to grasp the changing nature of water resources management and the urgent need for GCC economies to carefully manage food-water imported through supply chains. Secondly, the chapter analyses the options of GCC initiatives to invest in food supply chains to increase food supply through improved food-water management. The two case studies of investment opportunities and activities in Sudan and Romania are used to analyse the environmental risks and opportunities of wheat production. Wheat will be the crop of concern due to its social importance in Middle Eastern diets across the different social groups.

2. The Underlying Fundamentals of Food Supply Chains: Actors and Consequences

There are no water wars because food wars are not judged to be necessary. (Allan 2002)

The first section introduces the importance of food supply chains in understanding water security. It will highlight both the politicised relationships, as well as the inescapable bond, between sustainable food security and sustainable water security. This connection is important for public and private sector agents to recognise in order to improve water resources management hidden in supply chains. The relationship between water and food is exceptional. No other supply chain needs or consumes a natural resource in the proportions that the world's food supply chains use water resources consumptively. The water used to produce food will in this analysis be called food-water. The food supply chain arena looks superficially as a well-tuned metabolism (see Fig. 1), however, any analysis is meaningless without identifying the major players in the global food supply chain.

The food supply chain has a rich history that has led to what Friedman and McMichael (2009) have called *food regimes*. Food regime theory was developed by Friedman (1978, Friedman et al 1989) and McMichael (2009). It explains the underlying structures of the post-1950 structure of international agriculture. The 1939-1945 wars delivered an elemental shock to all the world's nations and particularly to the powerful warring parties. An unintended outcome was a strong post-war appetite for regulation. The "regulation of the food regime both underpinned and reflected changing balances of power among states, as well as between organised national lobbies and classes – farmers, workers, peasants – and capital" (McMichael 2009). Food regime analysis notes that the globalisation of modern agriculture first became evident with the British-led outsourcing of agricultural activities to colonies in tropical zones and then to former-colonies more generally. The first regime was associated with the wealth transferring and humanitarian disaster of the imperial tropical sugar trade of the 1750 to 1850 period. It was succeeded between about 1870 to 1930 by an era of grain and livestock production and export from settler colonies to industrialised and rapidly urbanizing European communities of all classes. This model functioned to feed the prospering middle classes around Europe with basic foodstuffs and with increasingly popular exotic commodities, such as tea and coffee. Interestingly most of the corporates now prominent in current global system had established themselves by 1870. The ABCD - global grain and livestock commodity traders (Murphy 2012), and two of the biggest food brands - Nestlé and Unilever, were already established by the late nineteenth century. They have been joined by a much more numerous group of long-established US corporations including, Pepsi, Coca-Cola and Kellogg since the beginning of the twentieth century.

The cereal traders, the ABCD corporates have been very significant actors in this third post 1980s Western-led global food system and its supply chains (Sojamo 2010, Sojamo

& Larson 2012, Murphy et al 2012). Another important feature of this third global food regime has been the emergence of a new supermarket retailer and wholesaler nexus. The very rapid expansion of this supermarket system involved an unprecedented rationalization of the food supply chains of the third food regime. For example the already impressive advanced weather and market information systems developed by the corporate food commodity traders was massively reinforced by the rapid evolution of computerised global data-handling across the regime.

Global food regimes have been around for much longer than the 20th century big-oil oligopoly. Interestingly the global food system is under-pinned by a group of staple-food suppliers that is much smaller than the current list of major oil and gas exporters. There are over twenty significant oil and gas producers accounting for most of the global oil and gas trade. In contrast the strategically important global grain trade is dominated by only five major net-exporters of water intensive staples - the United States, Canada, Brazil, Argentina and Australia. They trade with another very small number of grain commodity traders based in the US and France - the ABCD corporations, plus the Swiss-based Glencore, which is consolidating a number of smaller traders. They appear to operate a durable western-based global state/market alliance. But Keulertz (2012c) has pointed out that there is a new acronym to consider. The long established US/French aligned global ABCD phenomenon has been joined by an East Asian group of four global grain traders that aim to serve the needs and interests of mainly Asian net-grain importers. Three are based in Singapore and one in Indonesia. They are the NOWS corporations - the Noble Group, which grew by 25% in 2011, (Keulertz 2012c), Olam, Wilmar and Sinar Mas. In 2011 they handled just over 20% of the business transacted by that ABCD corporations but their trade is growing rapidly.

The global food regime could well be entering a new phase (Keulertz 2012b). The global alignments are likely to be subject to change. The ABCD traders have not yet adopted even the shared values vision of the brands such as Nestlé (Sojamo and Larson 2012). The NOWS corporations have no immediate incentive to adopt water value and water ecosystem aware systems. Although Olam is setting a principled natural resource awareness pace in its operations (Olam 2012).

However, the perceived "success" of the corporate food regime witnessed a severe crisis during the market volatilities of 2007/08 and 2010/11. Between 2006-2008, average rice prices rose by 217%, wheat by 136%, maize by 125% and soy by 107% (Murphy et al. 2012). This has exposed some dangerous features of the power asymmetries of the current global food regime, especially for GCC economies. However, before discussing the GCCs, it must be noted that many enduring sub-national food systems exist in regions with natural resources endowments readily available for farming purposes. These short food supply chains feed over 80% of the world's populations (Hoekstra et al 2012). Global food *trading* systems only ensure the food security of about 15% of the global population (Hoekstra et al 2012). This low proportion, however, belies its significance for regions such as the GCC where the farming potential is constrained due the hyper-arid climate. The successful servicing of this international demand for traded food, driven by food consumption in water and food deficit economies, keeps the world at peace – and the GCC economies food-secure. It must be emphasised that it is normal to live in a food deficit economy. About 160 out of the 210 economies in the world exist in conditions of inescapable food deficit of which they are usually innocent. There are no water wars because food wars are not judged to be necessary (Allan 2002).

There has been a clear shift in the nature of the global food system since the 1980s as a consequence of the expanded reach of global trans-national corporations based mainly in the United States and Europe. The prevailing third "corporate" food regime had unfolded to an Orwellian extent with very few actors now controlling global agricultural trade.

The commercial and communication competence of these long established and some new corporations has led to an unprecedented concentration of market power (Williams 2012). The transnational corporations in these food supply value chains - often referred to as the brands and the non-brands - operate across the world. They can operate in short sub-national supply chains. Increasingly influential, however, are the long global food supply value chains, which are very well integrated into the global food regime just discussed. These major players understand very well the operation of food supply chains and have well-developed information systems which uniquely privileges them in evaluating and handling environmental and market risks. They also have established - and sometimes still own - elements of the banking, hedging and insurance systems that underpin the operation of the supply chains. They stand out as the players who deploy huge influence.

3. The Water in Food Systems

Water used by society is often subject to a number of myths. First of all, human beings only require 3 litres per day for drinking purposes. In addition, we need approximately 300 litres for other domestic purposes such as cooking, personal hygiene and for electric appliances. The amount of water required for our daily use can easily be produced because it only amounts to approximately 10 per cent of the water on earth and these uses are not consumptive of water, only changing its quality, not its quantity. However, this chapter is dedicated to the remaining 90 per cent of water resources that is consumptively used in feeding us.

In order to improve the understanding of water resources of the general public, the Swedish scientist Malin Falkenmark introduced a chromatic distinction of water resources in the late 1980s (Falkenmark 1989). Falkenmark distinguished between several types of water. First, she highlighted the importance of green water. *Green water* is the water that – during a cropping season enabled by rainfall events – stays long enough in the root-zone to meet the evaporative needs of a harvestable crop. This water is immobile and not evenly distributed on earth. While temperate and tropical zones usually have an abundance of green water, semi-arid and arid zones such as the Gulf region have hardly any green water availability. Second, she identified blue water that is generated through the hydrological cycle. It is mainly "produced" through the runoff of precipitation to recharge rivers, streams, lakes and aquifers. This surface and groundwater is visible and can be pumped by engineers to provide drinking and sanitation water or, more importantly, irrigation water. The GCC economies almost entirely rely on blue water for domestic and agricultural use. Third, there is grey water. Grey water is recycled blue water that may have been used in irrigation but which did not evaporate and so returns to the blue-water system - usually carrying sediments and other impurities. It is widely used but proves to be expensive for agricultural production. Finally and increasingly important in the GCC, silver water is desalinated water generated from oceans (Haddadin 2008). This is by far the most expensive water option available to society. A desalination plant that provides drinking and sanitation water for 500,000 people can cost several billion US$. Much of the expense in blue, grey and silver water is associated with the energy costs of de-contaminating and transporting it to where it is needed. Producing silver water for agricultural production is a costly option that may even provide headaches to affluent economies such as those in the GCC. It is therefore no surprise the GCC economies have adopted a cheaper and so far reliable option to fill the hydrological gap: importing virtual water. Virtual water is a term that has been coined by Tony Allan in the 1990s. A group of researchers around Arjen Hoekstra in the Netherlands have quantified virtual water since the late 1990s. For instance, one kg of wheat requires approximately 1,300 litres of water during its production. One kg of beef even 16,000 litres if industrially produced in feedlots using animal feed from overseas. A central pivot in the provision of water is the food supply chain that ships water in food from one place in the world to another to achieve food security.

As a result, society's food supply chains utilise about 90% of the water used by society (Hoekstra et al 2012). We shall call this water in supply chains *food-water.* The other municipal water will be called *non-food water*. Of the massive proportion of natural green and blue water embedded in the world's food supply chains 90% is used in producing food and fibre on farms by farmers. Farms are clearly where improved returns to water are delivered and water ecosystems are stewarded.

The two types of natural water used by farmers are green water or blue water. Globally green water is the most important in terms of volume. About 80% of the food in global farming is produced with *green water* (Hoekstra et. al 2012). The metrics of the volumes of water available and of water transpired are both very difficult to capture (UNEP/GRID 2009) UNEP/GRID-Arendal (2009). Global estimates that 80 per cent of green water is used to produce society's food production are usually based on estimates of crop transpiration modelled from remotely sensed multispectral data. (Mulligan et al 2011, Mulligan 2013) These metrics can be also be augmented with estimates based on production data.

Green water has been very important throughout the 13 millennia of humanity's crop producing history. Despite the importance of green water in determining water and food security it is only very recently that its role has been recognised and metrics developed. Such metrics are still not included in national and international water datasets such as those of FAO, although there are moves to remedy this situation (Margat et al 2005).

Despite blue water being evident and very widely revered as a holy cultural presence it has rarely been valued as an economic input. Its value has not figured until its availability became scarce. Unfortunately by then it cannot be valued because that water resource use has become integral to unalterable livelihoods. These livelihoods are seen by society and by society's politics as unchangeable elements of an eternally politicised food economy.

The livelihoods that blue water has enabled are embedded in political economies created in a world that preceded the triple bottom-line assumptions of *people, profit* and *planet* (*The Economist* 2009) that were highlighted by activist scientists in the 1970s and 1980s. Adding the third environmental bottom-line is an elemental political challenge as it is asking society to address the second failure of capitalism. In the two centuries of industrialisation since 1800 – when society was being asked to address the first failure of capitalism – slavery and getting labour wrong, unprecedented demands for water have been imposed on blue water and the ecosystems, which rely on it. These new water demands were mainly a consequence of the increase in the human population from about one billion to the current seven billions.

4. Achieving Food-Water Security – the Middle Eastern Question

Farmers, Nature and don't waste food. (Japanese pre-meal vow/pledge)

The preceding sections have highlighted the global political economic dynamics of the food supply chain. If the "the world is out of water already", the Middle Eastern region is in the worst position of all. The "long twentieth century" in West Asia and North Africa involved very many tragic developments. Events in 1916, 1937, 1948, 1967, 1978, 1981, 1991, and 2003 had all a profound impact on agricultural history. The four main "bread baskets" in the region of the Fertile Crescent and Iran were almost continuously exposed to political changes that were markedly shaped by geopolitical games. The migration of millions of refugees in the wider Jordan and Euphrates and Tigris basins made agricultural development particularly difficult. Finally, the Cold War dichotomy led to different development strategies in the breadbasket countries with an availability of food water such as Egypt, Jordan, Palestine, Lebanon, Syria, Iraq and Iran. The result of the revolutions, wars, conflicts, socialist vs capitalist development path dependencies and high population growth is felt in agriculture. As of today, all major food water basins in the Middle East are in a state of political crisis or increasingly affected by climate change. None of the three major basins (the Nile Delta, the Jordan Basin and the Euphrates and Tigris) was able to develop according to the American Mid-West or the Brazilian Cerrado model, where a large part of the world's food water is now sourced.

Although the oil rent allowed Middle Eastern economies to develop some of their economic sectors at a breath-taking pace, the absence of food-water resources led to dependency on global supply chains (Klingbeil and Byiringiro 2013). The American trader Cargill for example estimates the MENA economies have been among their best customers since the end of the Second World War.

The GCC import water from other world regions (in particular from North America, Latin America, Oceania, Central Asia and Europe) of which they have no control apart from improved supply chain management.

Dependency on food water from the leading food regime companies active in the farming sectors of North America, South America, Australia and Eastern Europe is however a dangerous strategy. Instead a paradigm change - that values water and recognises the negative impacts of mis-managing it - is of great importance for the sustainable prosperity of the whole MENA region. The world needs new initiatives to promote food-water security. Starting to account for food water in supply chains faced by almost absolute water scarcity in the Middle East is the game changer proposed here.

It is argued first that, deploying a food water supply chain framework to understand how current food water losses can be avoided through a supply-chain wide recognition of the need for reporting and accounting rules that capture the value of water as an input as well as the long term costs to sustainability of mis-managing water ecosystems. Farmers, ag-traders, food processors and super-markets need to work with accountants and legislators to provide the existing food supply chain juggernauts with brakes and steering so that they stop destroying water environments. We are not suggesting a mechanistic installation of water pricing. The changes proposed will be very highly politicised. They have to be installed at a pace that is politically feasible introducing reporting and accounting rules shaped by the array of interests that populate food supply chains. There are huge contradictions to accommodate.

Second, improved storage and handling of raw produce through investment in the food supply chain can further increase water efficiency by cutting waste, especially in less-developed country supply-end settings. However, given rocketing population growth in the MENA region, these measures will not be enough. Therefore an investment strategy based on the food supply chain framework is needed. Ways must be identified that enable farmers and nature in the region, but mainly outside it, to make efficient use of food-water to meet the rising demand for food over the next forty years. The next section analyses two potential investment locations for wheat inside and outside the Arab world. By taking a hydrological perspective on how to increase agricultural production in a virgin-land area in Sudan and Romania, we look at the food supply chain from a bottom-up perspective. First and foremost we are interested in how to increase production, which is the starting point in our view for Middle Eastern supply chain food water management.

5. Two Case Studies (Sudan, Romania)

The WaterWorld model (v2.89) (Mulligan and Burke, 2005; Mulligan, 2012) was applied to understanding the impacts of proposed investments on baseline water quantity and quality for 10,000 ha investment in Sudan (Abu Hamad at 19° 32' 24.78"N, 33° 19' 14.12"E). A baseline water resources assessment simulation representing the mean climate 1950-2000 and land use at 2000 was run at 1-hectare spatial resolution, followed by a land cover and use change scenario representing conversion to wheat cover (0% tree, 90% herbaceous and 5% bare functional type per pixel) over clustered areas of approx 10,000 ha.

5.1. Sudan Case Study

The Sudan investment was focused near the Nile (at elevations less than 340 masl) as indicated in Fig. 1.

Stack 0 | Variable	Value
Form	landuse_change
Action	byft_by_suitability
Scenario name	test
Tree:	0
Herb:	90
Bare:	10
for area of land	10000
in units	ha
clustered	on
where	Elevation (SRTM Hydrosheds)
is	<
this value:	340
Land converted to	cropland
Intensity of use	1

Figure 1 **Set-up of WaterWorld Scenario for Sudan Investment**

Water balance (mm/yr) for the area was on average -810 with a 25th percentile of -850 and a 75th percentile of -780, an absolute minimum of -1,700 and maximum of -140. This reflects an area average precipitation (mm/yr) of 16 with an absolute minimum of 9 and maximum of 26. Actual evapo-transpiration (mm/yr) ranges from 170 to 1,800 with a mean of 830 with significant inputs of water from the Nile and from groundwater resources enabling evapotranspiration even under low rainfall, creating negative local water balances. Fog inputs are 0 mm/yr.

The impact of the applied scenario on water resources is an increase in actual evapotranspiration of 53 mm/yr on average for the converted areas. This led to a increase in evapotranspiration for the study area as a whole of 2.4 mm/yr (0.29 %) and no change in fog interception leading to an overall decrease in water balance of -2.4 mm/yr (0.3 %). The resulting decrease in water balance is highly variable depending on whether the new agriculture occurs on bare land or on land that was previously vegetated (Fig. 2) with previously bare land showing much greater losses upon conversion to agriculture.

Water quality (measured as the human footprint on water quality, Mulligan, 2009) decreases in the areas local to the investment but these effects propagate little downstream (Fig. 3) with values in most of the river courses increasing by a fraction of 1%.

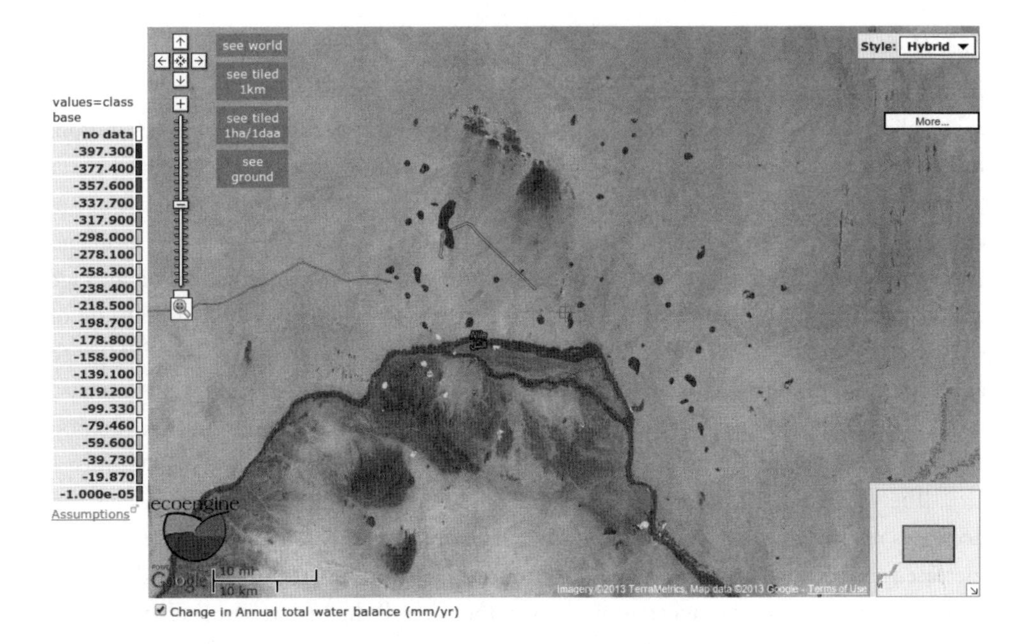

Figure 2 **Change in Annual Total Water Balance after Proposed Investment. Background image source: Google maps**

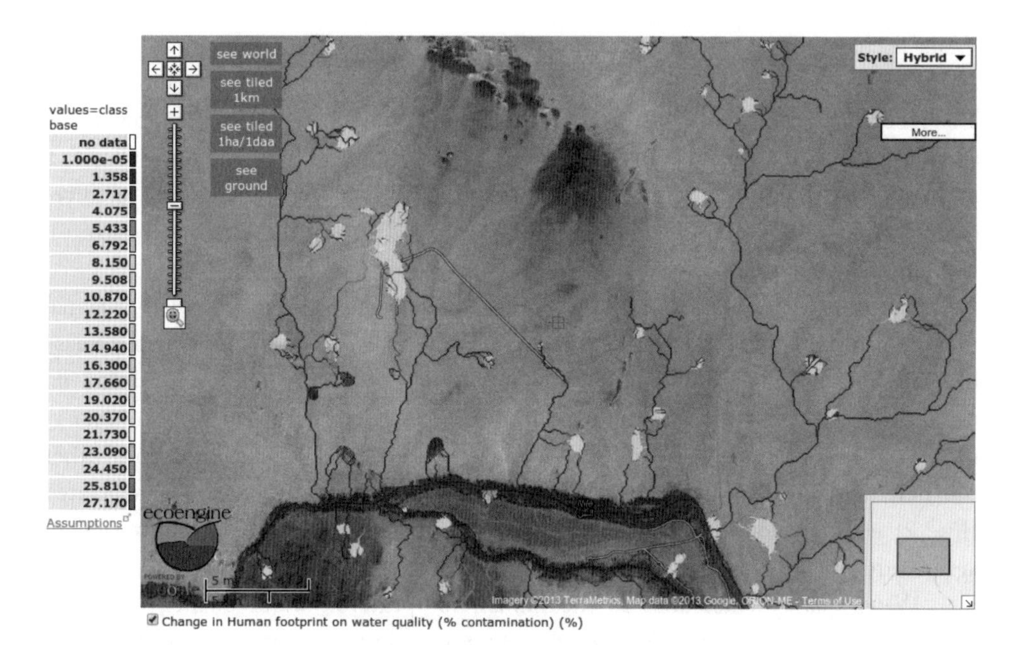

Figure 3 **Change in Water Quality after Proposed Investment. Background image source: Google maps**

5.2. Romania Case Study

In the case of Romania, a baseline water resources assessment simulation representing the mean climate 1950-2000 and land use at 2000 was run at 1-hectare spatial resolution, followed by a land cover and use change scenario representing conversion to wheat cover (0% tree, 90% herbaceous and 5% bare functional type per pixel) over clustered areas of approx. 10,000 ha. The Romania investment was focused on areas already under cropland (retasking them, see Fig. 4). Baseline water balance (mm/yr) for the area was on average 280 with a 25th percentile of 230 and a 75th percentile of 340 an absolute minimum of -370 and maximum of 790. This reflects an area average precipitation (mm/yr) of 360 with an absolute minimum of 0 and maximum of 500. Actual evapo-transpiration (mm/yr) ranges from 20 to 400 with a mean of 180. Fog inputs are low in relation to precipitation at 3.9 % on average, amounting to 17 mm/yr on average but ranging from 0 to 280 mm/yr.

The scenario led to increases in vegetation cover in some of the areas invested and decreases relative to prior vegetation in other areas. The areas with increased cover showed increased evapotranspiration of 5.6 mm/yr on average whereas the areas with decreased cover showing decreased evapotranspiration so that overall the investment leads to a small decrease in evapotranspiration of 0.65 mm/yr in all of the invested areas. This contributes to a mean decrease in water balance of -2.9 mm/yr in the invested areas since inputs of fog Impacts on water are significant in the areas invested, especially in the few areas in the north

Stack 0 | Variable	Value
Form	landuse_change
Action	byft_by_suitability
Scenario name	test
Tree:	0
Herb:	90
Bare:	10
for area of land	10000
in units	ha
clustered	on
where	Elevation (SRTM Hydrosheds)
is	<
this value:	340
Land converted to	cropland
Intensity of use	1

Figure 4 **WaterWorld Setup for the Romania Investment Scenario which Converts 10,000 ha of Existing Cropland to Wheat**

decrease(-2.9 mm/yr on average) in areas where the investments lead to tree cover loss. The overall effect is to decrease runoff in most areas, though downstream of the areas invested, these effects are very small (a fraction of 1%, see Fig. 5).

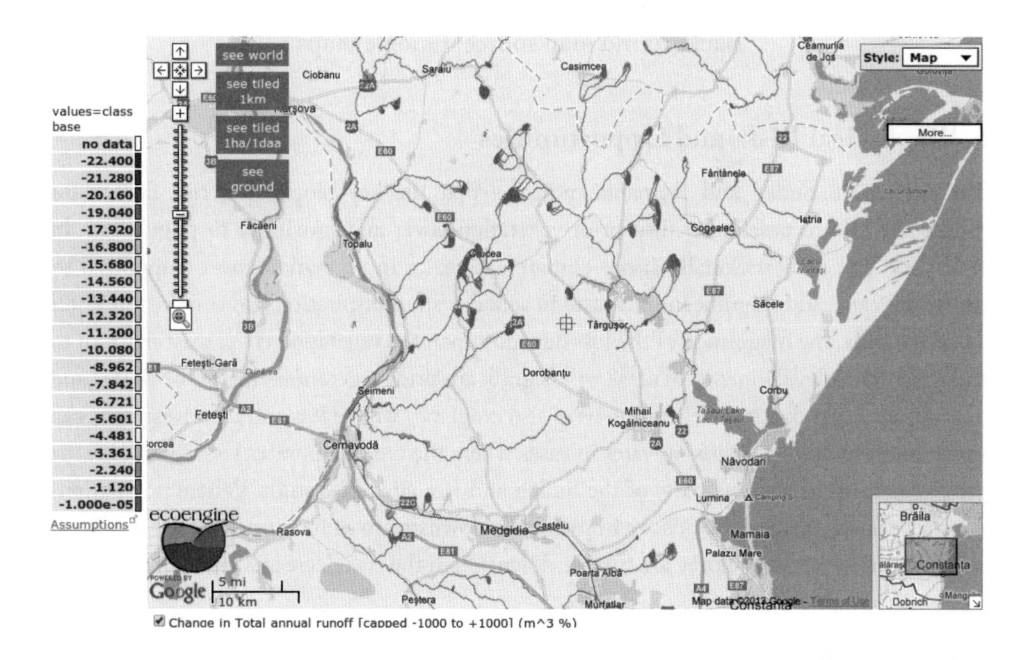

Figure 5 **Change in Annual Total Water Balance after Proposed Investment. Background map source: Google maps**

where forest is converted but again these propagate little downstream in this already heavily agriculturalised landscape (Fig. 6).

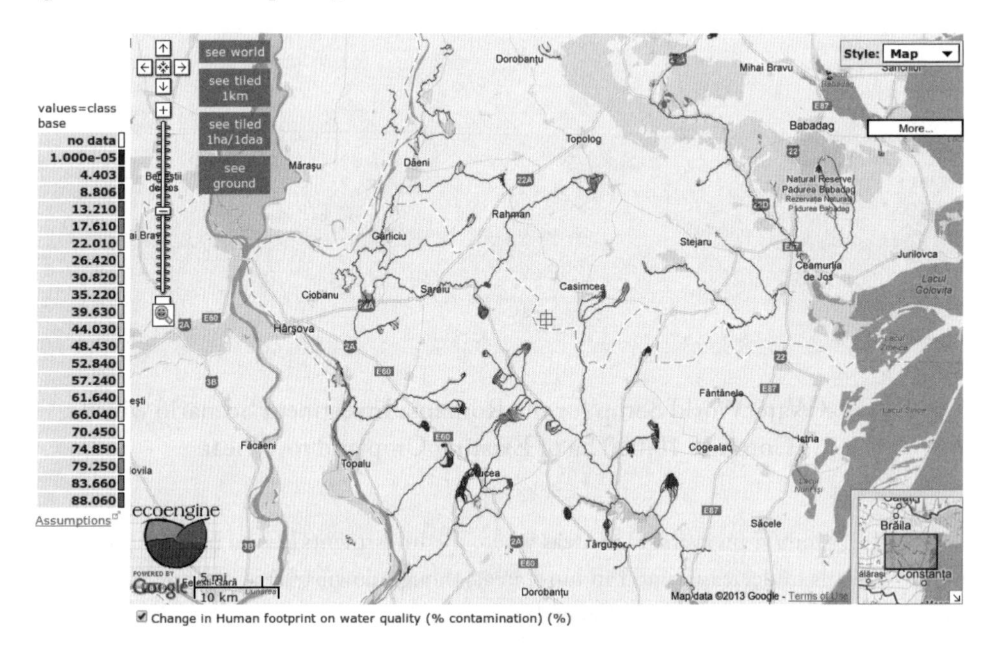

Figure 6 **Change in Water Quality after Proposed Investment. Background map source: Google maps**

6. Biophysical Risks and Opportunities

The analysis in Sudan and Romania indicates that the key biophysical risk factors for proposed FDI are baseline aridity, prior vegetation cover and proximity to population. In arid – and especially seasonally arid – climates, increases in vegetation cover (supported by irrigation) can lead to significant increases in actual evapotranspiration and thus loss in water balance locally. The magnitude of the loss depends upon the vegetation cover existing prior to the agricultural development. In areas with significant prior vegetation, the loss of water (and of water quality if the prior vegetation were also cropland) will be less and the investment can sometimes even lead to gains in water in cases where the crop evaporates less water than the prior vegetation as in some parts of the Sudan and Romania case study. Where populations or other existing water demands are immediately downstream of investments they may be at significant risk, but where they are further downstream the risk decreases greatly.

7. Discussion

Food water at the beginning of the supply chain is particularly at risk where virgin land is transformed into agricultural production without taking the local population and its livelihoods into account. If the area is selected for investment on the basis of factors that recognise local livelihoods, and in particular selected on the basis of prior vegetation (e.g. replacement of water-intensive cotton or lucerne crops), a transformation of the land to wheat production can turn investments into the desired win-win outcomes.

However, it is important to stress that the sound management of the ecosystem in Romania and Sudan can only be understood if the water balance of the investment is taken into account in relation to the production obtained. On average, the Sudanese wheat yield per hectare for the years 2008-2010 was at approximately 0.5 t/ha, which is significantly lower than the on average 7 t/ha achieved further upstream in Egypt. The average water footprint for wheat per hectare is 1830 m3 per tonne, hence in the Sudanese case study the water used to produce the 5000 tonnes of wheat is 9,300,000 m3 (Water Footprint 2013). With the same amount of water, the yields could be improved by investing further along the supply chain. First, through investment into fertilizers and pesticides produced in the GCC economies the yields could be drastically increased, if careful attention is paid to eco-efficient input. Second, the water losses associated with food spoilage of approximately 40 per cent due to poor storage and transportation could be minimised by increased investment into infrastructure and transportation facilities at the supply end. Third, consumer waste in GCC economies would have to minimised through improved consumer information or new legal frameworks taxing consumer waste.

In Romania, wheat harvests per hectare have been on average 3.5 tons from 2009-2011 (reference). The significantly higher number may signal a better investment climate in the EU state. However, similar risks akin to Sudan exist in Romania. Downstream water users may well be similarly affected if water is not valued from the supply end of the food supply chain. Although infrastructure in Romania is in a considerably better state than in Sudan, the competition for investment for GCC economies is by far greater than in Sudan. Romania has recently received a substantial interest from agribusiness companies to transform the former Communist country into a European "breadbasket"(Unteanu 2013). Western Europe is also affected by agricultural imports. At a time of harvest losses due to climate change, the Romanian option may sound appealing from an environmental and commercial perspective, yet the risks of long-term supply from an EU member state may be hard to estimate from a GCC investing economy. Thus, Romania's perceived agricultural potential may be a mirage in Eastern Europe. As a result, investment in Sudan against the commercial odds may be the most secure way of obtaining food security through investment in the bottom end of the food supply chain.

8. Conclusions

The results above may sound obvious from a business perspective. However, the importance of the role of food water in the food supply chain may help overcome food insecurity and place the farmer and livelihoods of farming communities at the heart of the decision-making process. Given the recent wave of activist literature against investment into farmland in Africa to avoid "land grabbing", the role of water provides two important lessons (LDPI 2010; Land Matrix 2013 among others). First, if existing farmland is selected compared to virgin land, the loss of water (quantity and quality) in the production process decreases significantly compared to a utilisation of unvegetated "virgin land", which is an expensive and potentially environmentally harmful activity. Replacing heavily vegetated natural land with agriculture will often lead to increases in water availability downstream because forests and shrublands generally consume much more water than agricultural land. However conversion of natural land to farmland has significant negative impacts on water quality downstream compared with utilising existing farmland and investing to improve productivity and reduce water use and water quality impacts. This also avoids the significant impacts on biodiversity and non-food ecosystem services that accrue from conversion of natural ecosystems to agriculture. In contrast to developing new lands, we propose to upscale existing agricultural production in Sudan by placing food water at the heart of the decision-making process. It is self-evident that such investment into existing agricultural schemes would require a number of governance challenges for a government such as Sudan in the target economies. From an investing economies' perspective this may involve a great number of political steps in uncertain territory. However, since food water is the main limiting factor in the GCC to upscale domestic agricultural production, it is worthwhile to understand the importance of food water all along the supply chain as the core input resource to overcome food insecurity and potentially hazardous scenarios for future economic growth in the GCC economies. After all, the world needs new initiatives. Placing food water at the heart of the investment objective; accounting for the water used and thus safeguarding the interests of all stakeholders could provide a new paradigm for improved supply chain management in the MENA region. It is a fundamental element of all the uncertain scenarios for the expanding GCC economies.

References

Allan, J. A. (2001) Virtual water: hydropolitics and the global economy. London: I B Tauris.

Allan, J. A. (2011) Virtual water: tackling the threat to the planets most precious resource. London: I B Tauris.

Allan, J. A. (2013) Food-water security: beyond water resources and the water sector. In Lankford, B., Bakker, K., Zeitoun, M. and Conway, D. Water security: principles, perspectives, practices. London: Earthscan.

Australian Government (2012) Australia's water. Canberra: Department of Sustainability, Environment, Water, Population and Communities.

DEFRA (2011) Future water: the Government's water strategy for England and Wales. London: DEFRA.

Douglas, M. (1992) Risk and blame: essays in cultural theory. London: New York: Routledge.

The Economist (2009) Triple Bottom Line. November 17th 2009. Print Edition. Available online: http://www.economist.com/node/14301663 (accessed 30 March 2014).

The Economist (2013) Daily Chart: How to feed a planet. http://www.economist.com/blogs/graphicdetail/2012/05/daily-chart-17 (accessed 31 March 2014).

ERD (2012) Confronting scarcity: managing water, energy and land. Brussels: European Development Report. http://www.erd-report.eu/erd/report_2011/report.htm (accessed 31 March 2014).

Falkenmark, M. (1986) Fresh water - time for a modified approach. Ambio, 15:4, pp 192-200.

FAO (2012) About the Voluntary Guidelines on the Responsible Governance of Tenure. Rome:FAO. Available online: http://www.fao.org/nr/tenure/voluntary-guidelines/en/ (accessed: 31 March 2014).

FAO (2013) The State of the World's Land and Water Resources for food and agriculture. Rome: FAO and Earthscan.

Federal Ministry of Economic Cooperation and Development of Germany (2006) Water sector strategy. Bonn: Federal Ministry of Economic Cooperation and Development.

Friedmann, H. (1978) World market, state and family farm: social bases of household production in an era of wage-labour. Comparative Studies in Society and History, 20(4). pp 545–86.

Friedmann, H. and McMichael, P. (1989) Agriculture and the state system: the rise and fall of national agricultures, 1870 to the present. Sociologia Ruralis, 29(2). pp 93–117.

FSDL (2012) Outcome Report of the 2012 Doha Conference on Food Security in Dry Lands. Doha: Qatar National Food Security Program.

The Guardian (2012) Energy company staff work at climate change ministry, The Guardian, 30 December 2012.

Hoekstra, A. Y. and Mekonnen, M. M. (2012) The Water Footprint of Humanity. PNAS. 109:9. pp 3232–3237.

Intelligence Community Assessment (2012) Global water security, A report requested by the US State Department, Washington DC: ICA.

IWMI (2010) Managing water for rainfed agriculture, IWMI Water Issue Brief, Colombo: IWMI.

Keulertz, M. (2012a) Land grabs and the green economy. In Allan, J. A., Keulertz, M., Sojamo, S. and Warner, J. Handbook of land and water grabs in Africa: foreign direct investment and food and water security. London: Routledge.

Keulertz, M. (2012b) The Middle Eastern Food Security Question and the Global Food Regime. Proceedings of the Conference on Food security in Global Dry Lands. Doha: Qatar Food Security Program.

Keulertz, M. (2012c) The Sudanese breadbasket: Land and water grabs by Middle Eastern Economies. Drivers of the Arab rush for land. Presentation at Oxford University 06 December 2012.

Keulertz, M. and Sojamo, S. (2013) Inverse globalization? The global agricultural trade system and Asian investments in African land and water resources. In Allan, J. A., Keulertz, M., Sojamo, S. and Warner, J. Handbook of land and water grabs in Africa: foreign direct investment and food and water security. London: Routledge.

Klingbeil, R. & Byiringiro, F. (2013) Food Security, Water Security, Improved Food Value Chains for Sustainable Socio-economic Development. Paper and Presentation at the SHARAKA conference "EU-GCC Regional Security Cooperation: Lessons Learned & Future Challenges", Qatar University, Doha, Qatar, 28-29 October 2013.

Land Matrix (2014) The Land Matrix. Rome: ILC. http://landportal.info/landmatrix (accessed 31 March 2014).

LDPI Initiative (2013) Land Deals Politics Initiative. Cape Town and Brighton: LDPI. http://www.plaas.org.za/ldpi (accessed 31 March 2014).

Margat, J., Franken, K and Fuarés, J-M (2005) Key water resources statistics in AQUASTAT: FAO's Global Information System on Water and Agriculture. Intersecretariat Working Group on Environment Statistics (IWG-Env). International Work Session on Water Statistics:Vienna. June 20-22 2005.

McKinsey (2012) McKinsey on sustainability & resource productivity. http://www.mckinsey.com/client_service/sustainability/latest_thinking/mckinsey_on_sustainability (accessed 31 March 2014).

McMichael, P. (2009) A food regime genealogy. Journal of Peasant Studies 36: 1. pp. 139-169.

Mulligan, M. (2009) The human water quality footprint: agricultural, industrial, and urban impacts on the quality of available water globally and in the Andean region. Proceedings of the International

Conference on Integrated Water Resource Management and Climate Change held at Cali, Colombia.
Mulligan, M. (2013) The water resource implications for and of FDI projects in Africa: a biophysical analysis of opportunity and risk. In Allan, J. A., Keulertz, M., Sojamo, S. and Warner, J. Handbook of land and water grabs in Africa: foreign direct investment and food and water security. London: Routledge.
Mulligan, M. (2013) WaterWorld: a self-parameterising, physically-based model for application in data-poor but problem-rich environments globally. Hydrology Research. In press.
Mulligan, M. and Burke, S.M. (2005) FIESTA: Fog Interception for the Enhancement of Streamflow in Tropical Areas. Report to UK DFID. http://www.ambiotek.com/fiesta (accessed 31 March 2014).
Mulligan, M., Saenz Cruz, L.L., Pena-Arancibia, J., Pandey, B., Mahé, G. and Fisher, M (2011) Water availability and use across the Challenge Program on Water and Food (CPWF) basins. Water International. 36: 1, 17-41.
Murphy, S., Burch, D. and Clapp, J. (2012) Cereal Secrets: the world's largest grain traders and global agriculture. Oxford: Oxfam.
OECD (2012a) OECD Environmental Outlook to 2050: The Consequences of Inaction. Paris: OECD.
OECD (2012b) Water Outlook to 2050: the OECD calls for early and strategic action. Global Water Forum: Marseilles 2011. http://www.globalwaterforum.org/2012/05/21/water-outlook-to-2050-the-oecd-calls-for-early-and-strategic-action/ (accessed 31 March 2014).
Olam (2012) Olam corporate responsibility report. Singapore: Olam.
Paalberg, R. (2010) Food politics: what everyone needs to know. Oxford: Oxford University Press.
Polman, P. (2011) The remedies for capitalism. McKinsey conversations with global leaders: Paul Polman of Unilever. San Francisco: McKinsey Quarterly. http://www.mckinsey.com/features/capitalism/paul_polman (accessed 31 March 2014).
Puma (2011) Combined Financial and Sustainability Report 2011. Herzogenaurach: Puma.
Reimer, J. J. (2012) On the economics of virtual water trade, Ecological Economics. Vol 75. pp 135–139.
Sojamo, S. (2010) Merchants of virtual water: the ABCD of agribusiness TNCs and global water security. Unpublished MSc dissertation. Department of Geography. King's College London.
Sojamo, S. and Larson, E.A. (2012) Investigating food and agribusiness corporations as global water security, management and governance agents: the case of Nestlé, Bunge and Cargill. Water Alternatives. 5(3): pp 619-635.
Stockholm International Water Institute and International Water Management Institute (2008) Saving water: from field to fork – curbing losses and wastage in the food chain. SIWI Policy Brief Stockholm: SIWI.
UNEP/GRID-Arendal (2009) Water Scarcity Index. Geneva: UNEP/GRID-Arendal Maps and Graphics Library. http://maps.grida.no/go/graphic/waterscarcity-index (accessed 31 March 2014).
UNESCO/UN-Water (2012) Managing water under uncertainty and risk. Paris: UNESCO.
UN Global Compact (2007) CEO Water Mandate. New York: United Nations. http://www.unglobalcompact.org/Issues/Environment/cEO_water_Mandate/index.html (accessed 31 March 2014).
United Nations Commodity Trade Statistics Database (2013) COMTRADE
database. New York: Department of Economic and Social Affairs/ Statistics
Division. http://comtrade.un.org/db/ (accessed 31 March 2014).
Unteanu, C. (2013) Could Romania be Europe's breadbasket. http://www.presseurop.eu/en/content/article/3242141-could-romania-be-europe-s-breadbasket?xtor=RSS-9 (accessed 31 March 2014).
URS (2013) Environmental data & hotspot impact research – Water metric feedback report: A report prepared for WRAP (Waste and Resources Action Programme, UK. Manchester: URS.
Waterwise (2007) Hidden Waters: A Waterwise Briefing by Joanne Zygmunt. London: Waterwise.
WBCSD (2006) Business in the world of water: WBCSD Water Scenarios to 2025. Geneva: WBCSD.
Williams, J (2012) Competition and efficiency in international food supply chains: improving food security. London: Routledge.
World Bank (2012) Addressing water scarcity in China. Washington, D.C.:World Bank 2012.

World Bank (2013) Cereal Yield per hectare. Washington, D.C.: World Bank. http://data.worldbank.org/indicator/AG.YLD.CREL.KG (accessed 31 March 2014).
World Business Council for Sustainable Development (2012) Global water footprint tool. www.ceowatermandate.org (accessed 31 March 2014).
World Economic Forum (2011) Water security: the water-energy-food-climate nexus. Washington D.C.: Island Press.
WRAP (2011) New estimates for household food waste in the UK. London: WRAP.
WWF (2008) UK Water Footprint: the impact of the UK's food and fibre consumption on global water resources. Leatherhead: WWF.
WWF (2012) Assessing water risk: a practical approach for financial institutions. Leatherhead: WWF.

10

The Use of Participatory Methods & Simulation Tools to Understand the Complexity of Rural Food Security

Samantha Dobbie, James G. Dyke and Kate Schreckenberg

Introduction

Food security is a multidimensional phenomenon. According to the FAO (1996), it occurs "when all people, at all times have physical and economic access to sufficient, safe and nutritious food to meet their dietary needs and food preferences for an active and healthy life". Within Arab countries, a number of contributing factors act to undermine the realization of food security. Food imports are responsible for over 50 percent of calories consumed (FAO, 2008). However, the capacity to generate foreign exchange to finance such food imports is dwindling (Breisinger *et al.*, 2012). The global food price spikes of 2008 and 2011 highlighted the vulnerability of Arab countries to swings in commodity prices and has lead Haddad *et al.* (2011) to call for a renewed focus upon domestic and regional food production.

A number of constraints associated with agriculture in less developed countries are shared by Arab nations. High population densities drive small plot sizes, while poor soil quality and a dependence upon rain fed agriculture act to escalate vulnerability to climatic shocks (Sahley *et al.*, 2005). In Sudan and Yemen for instance, rainfed agriculture accounts for over 80 percent of cereal production (FAO, 2008). In general, low and poorly distributed rainfall throughout the Arab region (250 – 600 mm) leads to frequent dry spells and moisture stress (Haddad *et al.* 2011).

In response to drought, rural households exhibit a number of coping strategies. A survey of over 500 individuals in Western Sudan uncovered alterations to consumption practices in response to shocks (Ibnouf, 2011). The number and diversity of meals consumed by a household for example, were frequently reported to have been reduced (Ibnouf, 2011). In addition, results revealed changes to cropping patterns with increased cultivation of drought resistant crops such as sorghum and millet (Ibnouf, 2011). Further studies have found households in drought prone areas are often reliant upon social networks and safety nets in the form of input subsidies and food for work programs (Ziervogel *et al.*, 2006).

1.1. The Potential for Simulation Tools

Attempts have been made to employ modelling techniques in evaluating strategies that promote food security among smallholder households. Masters *et al.* (2000) for instance define an optimization model that interprets the relationship between household nutritional intake and labour productivity, with added consumption, resource and borrowing constraints. The model is calibrated with household survey data for Malawi and employed to investigate two broad policies, including: the input-market liberalisation program for tobacco and other crops, as well as implementation of the Starter Pack Scheme for input subsidies (Master *et al.*, 2000). Dorward (2003) on the other hand, documents the development of mathematical farm/household models to explore responses to change in the form of maize price increases and fertiliser use. Whilst Thangata *et al.* (2002), describe the use of a dynamic mathematical model to decipher drivers of agroforestry adoption and household decision-making.

It is the complex social-ecological nature of smallholder dynamics, which makes ABM an appropriate tool for studying food security (Mena *et al.*, 2011). This technique consists of a computerised simulation of agents located within an environment, which interact through predisposed rules (Farmer & Foley, 2009). Behaviour at the system level is an emergent property of the collective behaviour of individual agents at the local level (Matthews *et al.*, 2007). A number of studies have employed ABMs to investigate aspects of social-ecological systems, for example: models of resource management, land use change and spread of innovations (Schlüter and Pahl-Wostl, 2007; Deadman, *et al.*, 2004; Johnson *et al.*, 2006).

Depending on the type of rules they abide by, agent-based models can be categorised as abstract, experimental, historical or empirical (Berger & Schreinemachers, 2006). In this study, attention is given to empirical agent-based models. Here rules that govern the behaviour of agents and the environment are centred upon empirical observation. Such models offer the potential to explore the effect of real-world policies and investigate scenarios, providing a boundary object at the interface between science and policy (Matthews *et al.*, 2007; Parker *et al.*, 2003). A study by Holtz & Pahl-Wostl (2011) for example, employed an empirical ABM to shed light upon groundwater over-exploitation in the Upper Guadiana, Spain. Different farmer types, part-time, family-based and business orientated, were found to exhibit different responses to changing policies and available technologies (Holtz & Pahl-Wostl, 2011).

According to Berger *et* al. (2006) empirical agent-based models may assist with the targeting of policy and extension programs, especially within less developed countries. Often policy makers and agricultural extension program designers are faced with a trade-off between avoiding a blanket approach to the development of policies and programs, while acknowledging it is impossible to take into account each individuals circumstances (Bohnet *et al.*, 2011). By utilising the multi-scalar nature of ABM, it may be possible to construct policies based upon regional, or district level behaviour that emerges as a result of local interactions at the household level (Berger *et al.*, 2006).

An (2012) however, argues that a shortage of effective architectures and protocols to represent agents and their interactions may limit the capacity of ABM to describe social ecological systems accurately. As a result, the ability of ABM to inform credible environmental decisions may be constrained (Le *et al.*, 2011). Furthermore, the construction and validation of empirical ABM is determined by data generated by fieldwork rather than *a priori* assumptions (Janssen & Ostrom, 2006). Therefore, the potential to design and parameterise robust empirical models is limited by the ability to generate sufficient data (Smajgl *et al.*, 2011).

A number of different approaches to empirical data collection are documented throughout the literature (Robinson, 2007; Valbuena et al., 2008; Smajgl et al., 2011). According to Robinson (2007) these can be grouped into five different categories: i) sample surveys; ii) participant observation; iii) field and laboratory experiments; iv) companion modelling and v) GIS and remotely sensed spatial data; applicability of the different approaches is governed by the ABM in mind.

1.2. The Use of Participatory Tools

For this study, attention was given to participatory rural appraisal (PRA) exercises. PRA is a broad term offered to a range of techniques and methods for field-based data collection, pioneered by the international development community. It encompasses a growing range of applied approaches including: matrix scoring, seasonal calendars, wellbeing ranking and mapping (Chambers, 1994). More recently, PRA exercises have been designed and implemented to aid the construction of empirical models (Krywkow *et al.*, 2012).

Frameworks have also been described which provide guidelines for the systematic creation of ABM. Smajgl *et al.* (2011) for example, outline a step-by-step procedure for the empirical characterisation of agent behaviours. This starts with the identification of agent behavioural categories, which are then scaled to represent the whole population of agents (Smajgl *et al.*, 2011). Methods for the formation of agent behavioural categories, termed agent types herein, are expanded upon by Valbuena *et al.* (2008), while techniques to generate populations of different behavioural categories are provided by Berger & Schreinemachers (2006).

The aim of this study was to determine whether an empirical ABM of rural households could be created in a systematic manner to represent social ecological interactions at a level relevant to policy analysis. Is it possible to use results from participatory exercises in the characterisation of agent types? Once constructed, can these agent types be scaled effectively? And what is the potential for such simulation tools to elicit greater understanding of the complexity of rural food security?

2. Methodology

The methodology draws from two existing frameworks for empirical ABM (Valbuena *et al*, 2008; Smajgl *et al.*, 2011). Three components can be recognised: the design of a PRA exercise, the construction of an agent based model and model implementation (Figure 1). Feedbacks exist between the first two components, with the definition of agent types informing the design of the PRA exercise and results of the PRA exercises dictating agent type parameterisation. The remainder of this section will consider key elements of the methodology in more detail. Although construction of the ABM and design of the PRA exercise is not considered to be a mutually exclusive process, for the sake of clarity they will be discussed separately.

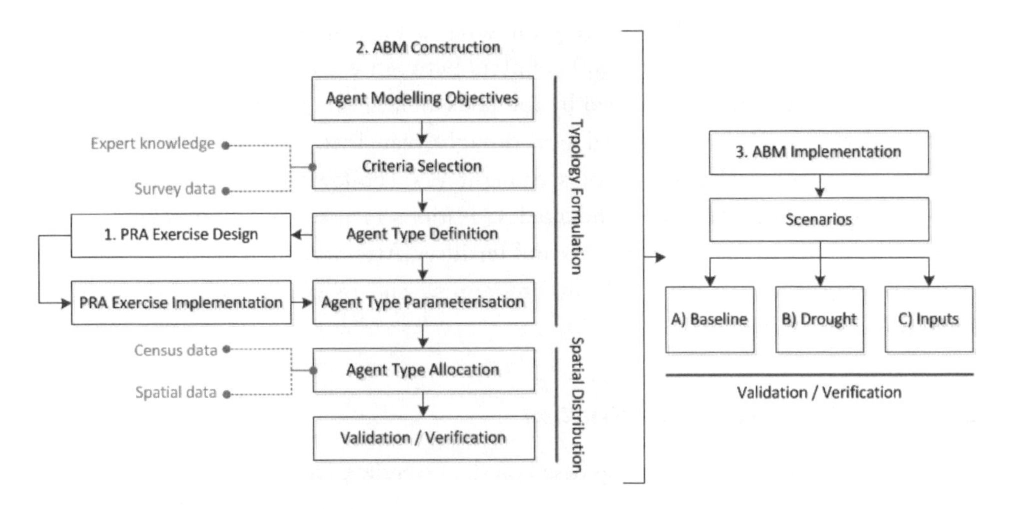

Figure 1 **Outline of Methodology; Empirical Construction of the ABM draws from two existing Frameworks described by Valbuena et al. (2008) and Smajgl et al. (2011).**

Typology formation acts to identify and characterize key household types present within the study area. While spatial distribution describes the process of allocating a household type to a population of model agents. PRA refers to Participatory Rural Appraisal.

2.1. PRA Exercise Design & Implementation

A Participatory Rural Appraisal (PRA) exercise was developed in order to explore i) the impact of drought upon household decision-making and ii) the impact of input subsidies upon household food security. In order to achieve this, the PRA was composed of two tasks: firstly a seasonal ranking exercise and secondly, a set of interview questions on input subsidies.

The study area comprised of two Traditional Authorities located within Southern Malawi. A total of four different rural villages were visited, referred to as village 1, 2, 3 and 4 herein. Research outlined within this report was conducted within the context of a longer-term relationship with the community as the villages selected for this study are part of a large multidisciplinary project, Attaining Sustainable Services from Ecosystems (ASSETS).[1] Agent types defined in the early stages of model development were employed to target four distinct groups of smallholders, based upon gender and wellbeing.

Prior to data collection, results from ASSETS Seasonal Calendar and Coping Strategy exercises were used to identify 14 household activities (Schreckenberg *et al.*, 2012). During the first part of the PRA exercise villagers were given the opportunity to identify an additional activity, bringing the total to 15 in some cases. For each month participants were asked to divide 60 counters between each activity, reflective of the amount of time and effort bestowed by the household. Participants were first asked to consider decisions made in a typical, non-drought year. Then the exercise was repeated to take into account the impact of a drought year. Following the seasonal ranking exercise, a group interview was conducted. Questions focussed upon household access to input subsidies in the form of seeds and fertiliser. Attempts were made to gauge the perceived impact of the input subsidy programme and its current role in promoting household food security.

2.2. Agent-based Model Construction

Table 1 provides a model description based on the Overview, Design concepts and Details (ODD) protocol (Grimm *et al.* 2006, 2010). The agent-based model was built in NetLogo 5.0.4, with construction involving a number of steps (See Figure 1). The remainder of this section describes the formation and allocation of agent types in greater detail. An overview of model procedures and parameters can be found in the Appendix. The model code is published in full online. [2]

2.2.1. Formation of Agent Types

Gender and wellbeing were selected as the two criteria for defining agent types. A total of 4 different agent types were identified, these comprised:

- Type 1: Male Heads of Household of medium or rich wellbeing
- Type 2: Male Heads of Household of poor or very poor wellbeing
- Type 3: Female Heads of Household of medium or rich wellbeing
- Type 4: Female Heads of Household of poor or very poor wellbeing

Table 1 **Description of the ABM based on the Overview, Design Concepts and Details (ODD) protocol (Grimm *et al.*, 2006).**

Model Description

Purpose: The purpose of the agent-based model is to simulate the impact of drought and input subsidies upon the livelihoods of rural households.
State variables and scales: Households are a main entity, which are simulated with attributes land area (ha), livestock, poultry and food adequacy (see Table 3). Households can be one of four agent types based upon gender and wellbeing (see Section 2.2.1.) A total of 15 808 households simulated to represent the population of two Traditional Authorities within Southern Malawi. The model is non-spatial.
Process overview and scheduling: The model works in monthly time steps. First, drought occurs according to historic climate data. Households then decide how much effort to bestow on productive activities including: farming maize, tending livestock and selling or buying at the market etc. Decisions are affected by the month, whether there is a drought (or not) and access to input subsidies. Household wealth and food adequacy can then be calculated.
Design concept: Emergent phenomena include the level of wellbeing. Households are reactive and do not attempt to optimise decisions or adapt over time.
Initialisation: The incidence of drought is initialised with historic climate data. Characterisation of household types is described in Section 2.2.1. and 2.2.2.
Input and sub-models: Parameter values build on primary data and expert knowledge. See Appendix for details.

Table 2 outlines key characteristics associated with each of the agent types. These were identified using results from ASSETS Wellbeing Ranking exercises already conducted in the villages of interest. Definitions of the agent types were used to target participants for the PRA exercises. The outcomes of PRA exercises were then used to construct behavioural rules for each of the agent types.

For each month, across each of the four villages, means and standard deviations were calculated for the proportion of time spent on each of the activities. Results were used to construct log-normal distributions for each of the activities according to the month and agent type. Log-normal distributions were chosen to provide only positive values, to reflect the fact that smallholders have two options: i) to allocate effort to a given activity that month, or ii) not to allocate any effort. Behavioural rules are summarized as in Table 3, for each of the 14 productive activities a total of 96 combinations exist, this is based upon whether it is a drought year, the agent type and the month in question.

2.2.2. Allocation of Agent Types

The model was to be set at a level relevant for policy and scenario analysis. Within Malawi local government is conducted at the district level, administered by district commissioners, who are appointed by the central local government (Njaya, 2007). Districts are subdivided into Traditional Authorities (TAs) overseen by chiefs, who play an integral role in local government politics. Traditional authority structures are a legacy of the colonial era. Introduced by English colonialists in the 1940s, indirect rule gave chiefs the responsibility to collect taxes, fees and dues (Njaya, 2007). Following independence, the Chiefs Act (1967: Section 7) granted traditional leaders a role in settling disputes and allocating customary land. According to Njaya (2007) a number of development projects have also depended upon the support of traditional leaders. As a result, this study focused upon two neighbouring Traditional Authorities within Southern Malawi.

Table 2 **Common Characteristics of the four Agent Types selected based on the two Criteria: Gender and Wellbeing. Characteristics are drawn from Results of ASSETS wellbeing Ranking Exercises already conducted within the Villages of Interest.**

		Perceived Wellbeing	
		Poor & Very Poor	Medium & Rich
Gender of Household Head	Female/ Male	< 1.0 ha of cultivated land Own poultry only Inadequate food availability for the year Access to public healthcare only	> 1.0 ha of cultivated land Own livestock including goats and poultry Adequate food availability for the year Access to both private and public healthcare

Table 3 **Example of a Behavioural Rule for the Allocation of Effort between Productive Activities; x can be either Agent Type 1, 2, 3 or 4; y can be Jan, Feb, Mar etc. and z is one of the 14 Productive Actives e.g. Maize Farming. SD is Standard Deviation.**

Behavioural rules
If drought is FALSE and agent-type is x and month is y then probability of activity z is log-normal mean SD
If drought is TRUE and agent-type is x and month is y then probability of activity z is log-normal mean SD

To ensure the ABM was representative of the study area, scaling of the four agent types was required. PRA exercises had been conducted once for each of the agent types in four different villages throughout the study area. The total sample size was 16, with 4 replicates for each of the agent types. In order to build a model representative of the study area, 15 808 households known to exist within the two Traditional Authorities were to be allocated with an agent type (NSO, 2012). In accordance with Smajgl *et al.* (2011), as the PRA exercise had targeted the four household types in an unrepresentative manner this required the use of household survey data.

The most recent integrated household survey for Malawi, IHS3, was conducted in 2010-11. Of the 12 000 households sampled for the IHS3, just 48 resided within the two Traditional Authorities of interest, 3 of which had to be discounted due to insufficient data collection (NSO, 2012). When complete data sets are unavailable, Berger & Schreinemachers (2006) advocate the use of random assignment. The impact of uncertainty upon simulation outcomes can then be judged by testing sensitivity to repeated random assignments (Berger & Schreinemachers, 2006). By creating a number of random agent populations and investigating the effect this has on model outcomes, it is possible to gauge the effect of random assignment. For this study, generation of the agent population relied upon two stages: firstly, the identification of four clusters within the survey data that corresponded to the four agent types and secondly, the generation of 15 808 households based upon the survey data using Monte Carlo techniques (Berger & Schreinemachers, 2006).

A k-means cluster analysis enabled the identification of four clusters within the sampled IHS3 survey data (n = 45) (see Table 4). Variables used for the analysis were selected based upon the common characteristics of the four agent types, identified from ASSETS wellbeing ranking exercises (See Table 2). Variables filtered from the survey data included: the area of land owned in ha, the number of livestock (including goats and sheep), the number of poultry, health care and a self-assessment of food adequacy (See Table 2).

Monte Carlo techniques were then employed to generate the entire agent population of 15 808 households (Berger & Schreinemachers, 2006). An empirical cumulative distribution function (CDF) was first created at the population level to determine agent type (Berger & Schreinemachers, 2006). A random integer between 0 and 100 was drawn for each agent and the agent type read from the y-axis (see Figure 2).Conducting this procedure for all 15 808 agents acted to recreate the empirical distribution. This process was then repeated at the level of individual clusters (or agent types) to allocate resource endowments. Each agent was assigned a type, an area of land, number of livestock, number of poultry and food-adequacy value. Consistency checks were then undertaken at the population, cluster and individual level to ensure the generated agents were statistically consistent and realistic (Berger & Schreinemachers, 2006).

Table 4 **Results of a k-means Cluster Analysis using Data from the IHS3 corresponding to the two Traditional Authorities of Interest (n = 45). Clusters 1 to 4 correspond to Agent Types 1 to 4, with Cluster 1 representing Male Heads of Household of medium and rich wellbeing etc.**

	Gender of Household Head			
	Male		Female	
Cluster	1	2	3	4
Area of land (ha)	0.96	0.74	0.65	0.22
No. of livestock	4	0	4	0
No. of poultry	15	0	15	0
Standard of health care received (1 is less than adequate, 2 is just adequate and 3 is more than adequate)	2	2	2	2
Food adequacy (1 is less than adequate, 2 is just adequate and 3 is more than adequate)	2	2	2	1
Proportion of sample population (%)	28.9	48.9	4.4	1.8

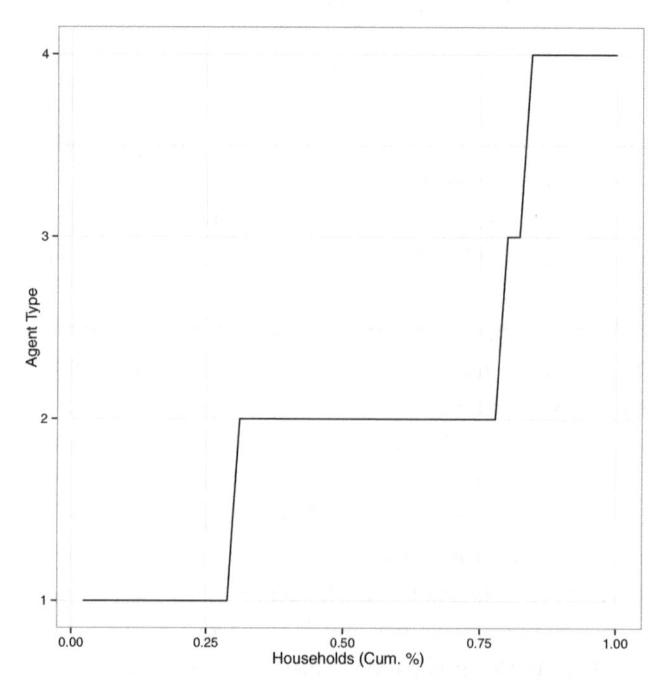

Figure 2 **Empirical Cumulative Distribution of Agent Types over all Households in the Sample**

At the population level average resource endowments for the population had to lie within the confidence intervals of each estimated sample mean. While at the cluster level, resource endowments for the generated agents needed to correlate with clusters identified within the IHS3 sample data set (n = 45). Finally, at the individual level, correlations between endowments, such as land and labour needed to be reproduced. Results of consistency checks are outlined within Section 3.2.

2.3. Model Implementation

Once typology formation and allocation had been fulfilled, the model could be implemented. Baseline, drought and input subsidy scenarios were used to test sensitivity of model parameters (See Appendix for details).

3. Results

3.1. PRA Implementation

Four PRA exercises were undertaken in each village, corresponding to each of the four target groups. Three distinct seasons were identified and agreed upon by each of the villages: a rainy season lasting from November to March, a cold season from April to July and a hot season from August to October. Annually, four household activities were found to dominate, including: buying from the market, collecting fire wood, caring for livestock, and maize farming. A one-way analysis of variance confirmed significant differences between the proportion of time spent on activities in baseline and drought years; for example, maize farming ($F_{1,382} = 31.8$, $p < 3.3\ e^{-08}$), tending to livestock ($F_{1,382} = 12.9$, $p < 0.0004$) and ganyu ($F_{1,382} = 10.9$, $p < 0.001$) all showed significant differences. Common coping strategies also included a slight increase in the growth of cassava, which villagers regarded to be more drought tolerant than maize.

For all villages, reliance on the market was found to increase dramatically during drought years; on average the proportion of time spent buying goods from the market rose by 14 per cent. In each of the PRA exercises, participants made it clear that they would rather cultivate crops such as cassava and sweet potato, fetch fire wood and collect wild and indigenous fruits to be sold at market in exchange for more maize, rather than for consumption.

Despite such similarities, distinctions between each of the villages were also apparent. Village 3 for example, neglected livestock in times of drought, while other villages invested more time in a bid to increase market value. In order to investigate inter-village differences further, for each activity a one-way analysis of variance was conducted. Multiple comparisons (Tukey's HSD, overall alpha level = 0.01) revealed significant two-way contrasts between the four villages throughout both baseline and drought years (Table 5). Such variance between the villages could undermine the formation and allocation of agent types (See Section 3.2.).

All of the participants interviewed reported access to input subsidies, governed by vouchers. No differences were found between the quantities of inputs provided. In all four villages, coupons were exchanged for 50kg bags of maize fertiliser, 5kg of maize seed and 2kg of legume seeds. Input subsidies were clearly dominated by maize, only men from village 3 of poor and very poor wellbeing acknowledged access to tobacco fertiliser and rice seed. Little difference was found regarding the months that input subsidies became available. Participants from all four villages stated that in a typical year, vouchers were available between the months of September to December.

Table 5 One-Way Analysis of Variance for each of the Household Activities to determine the Effect of Village Membership upon the Proportion of Time allocated. Tukey's HSD permits multiple comparisons with an overall alpha level of 0.01; * denote significant village two-way contrasts. The analysis uses results from village 3-4 for both baseline and drought years.

Household activity	One-way ANOVA	Tukey's HSD Significant village two-way contrasts					
		1-2	1-3	1-4	2-3	2-4	3-4
Maize farming	$F_{3,380} = 3.79, p = 0.01$				*		
Sweet potato farming	$F_{3,380} = 85.4, p < 0.01$			*		*	*
Rice farming	$F_{3,380} = 148.2, p < 0.01$	*	*	*		*	*
Pigeon Pea farming	$F_{3,380} = 191.6, p < 0.01$			*		*	*
Tobacco growing	$F_{3,380} = 24.4, p < 0.01$	*	*	*		*	*
Cassava farming	$F_{3,380} = 26.5, p < 0.01$	*		*	*		*
Fishing	$F_{3,380} = 13.2, p < 0.01$	*	*	*		*	
Hunting	$F_{3,380} = 25.7, p < 0.01$	*		*	*		
Tending to livestock	$F_{3,380} = 7.6, p < 0.01$			*		*	*
Collecting wild food/ indigenous fruits	$F_{3,380} = 6.1, p < 0.01$			*		*	*
Collecting fire wood	$F_{3,380} = 16.3, p < 0.01$	*		*	*		*
Ganyu	$F_{3,380} = 20.4, p < 0.01$		*	*	*		*
Sell at market	$F_{3,380} = 1.4, p > 0.01$						
Buy from market	$F_{3,380} = 31.1, p < 0.01$			*	*	*	*

3.2. Formation & Allocation of Agent Types

A total of four clusters were identified within the sample IHS3 data (See Table 4). Monte Carlo techniques were then employed to generate a population of 15 800 households. Validation and verification of the generated agent population occurred at the population, cluster and individual level (Berger & Schreinemachers, 2006). At the population level, resource endowment of the agent population had to lie within the confidence intervals of each estimated sample mean. Results from an agent population generated 100 times using a different random seed value can be compared with IHS3 data (Table 6). Results confirm that the sample population was well replicated at the population level.

At the cluster level, the correlation matrix of resource endowments for each of the four generated agent types has to reflect that of the IHS3 sample population (n=45). Two box plots compare the distribution of land area owned and perceived food adequacy between the four typologies extracted from survey data and generated using Monte Carlo techniques (Figure 3). The Figure shows that for male headed households (type 1 and type 2), median values do not differ much between the survey population and the agent population. For female headed households (type 3 and type 4) however, median values for land and adequacy do show differences. This is most likely due to a scarcity of survey data for female headed households. Sample sizes comprise just 2 and 8 households for agent type 3 and 4, respectively.

Table 6 **Resource Endowments of the Survey Population compared to meta-averages of the Agent Population. Agent population comprises the average of 100 agent populations, generated using a different random seed value each time. Standard error (SE) equates to the SE of the within survey population average. Standard deviation (SD) refers to the SD of the average across agent populations.**

	Population	Average	Standard Error/ Standard Deviation	Confidence Interval	
Agent Type					
	Survey	2.1	0.15	1.8	2.4
	Agent	2.1	0.01		
Resource					
Land (ha)	*Survey*	0.33	0.03	0.27	0.40
	Agent	0.34	0.00		
Livestock	*Survey*	1.3	0.27	0.8	1.9
	Agent	1.3	0.01		
Poultry	*Survey*	5.4	1.1	3.2	7.6
	Agent	5.4	0.01		
Food adequacy	*Survey*	1.6	0.08	1.4	1.8
	Agent	1.6	0.04		

At the individual level, the success of the approach in reproducing correlations between two particular resource endowments: area of land and number of livestock owned can be illustrated (Figure 4). Figure 4 illustrates that correlation between land area and number of livestock owned as observed in the survey data is well replicated in the agent populations.

Despite consistencies at the population, cluster and individual level, inter-village differences identified within Table 5, undermine the formation and allocation of agent types. Variance between the villages coupled with small sample sizes hinders the ability to make robust inferences concerning meta-village populations. Fundamentally, this limits the studies impact on policy making at regional level as it is not possible to make robust inferences about collective actions of large numbers of potentially different village response to different scenarios.

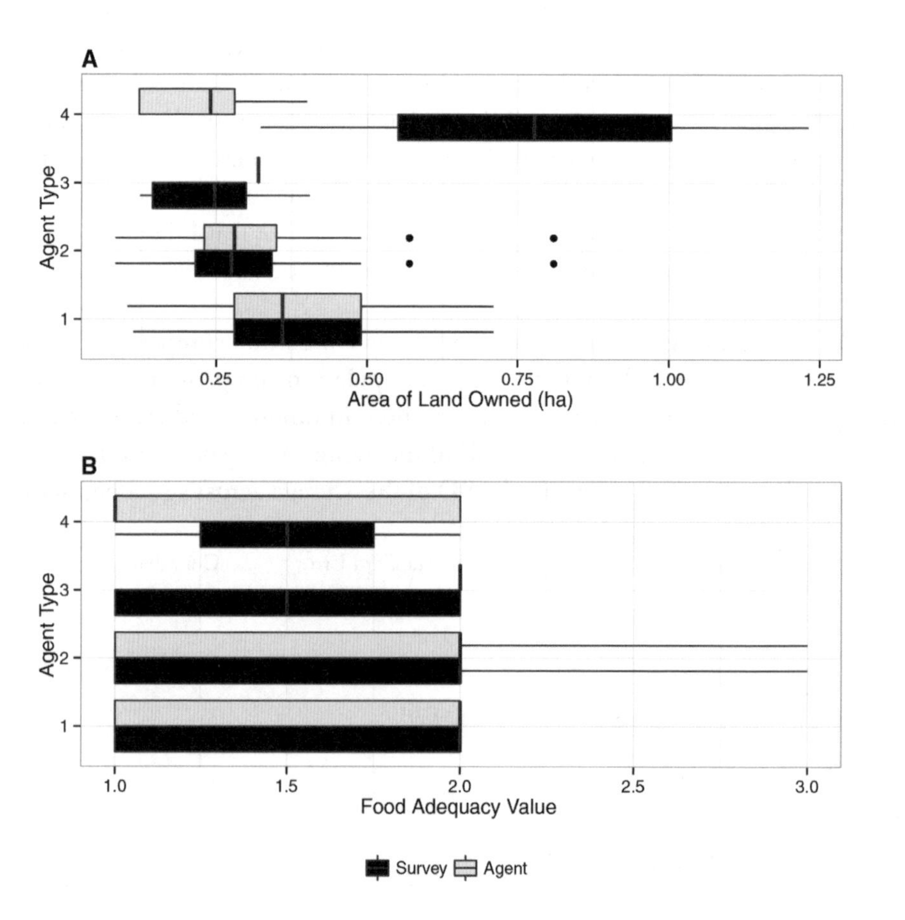

***Figure 3* Boxplots for the distribution of the area of land owned and food adequacy value over clusters. Agent types are defined as 1: male Heads of Household (HH) of medium or rich wellbeing, 2: male HH of poor or very poor wellbeing, 3: female HH of medium or rich wellbeing and 4: female HH of poor or very poor wellbeing.**

3.3. Model Implementation

Simulations followed three different scenarios outlined earlier, namely, baseline, drought and input subsidies. Differences are revealed between the behavioural decisions of agents separated by type. Women of poor and very poor wellbeing (type 4) for instance, allocate less time to maize cultivation when compared with other agent types, instead they focus on the collection of firewood and growth of cassava.

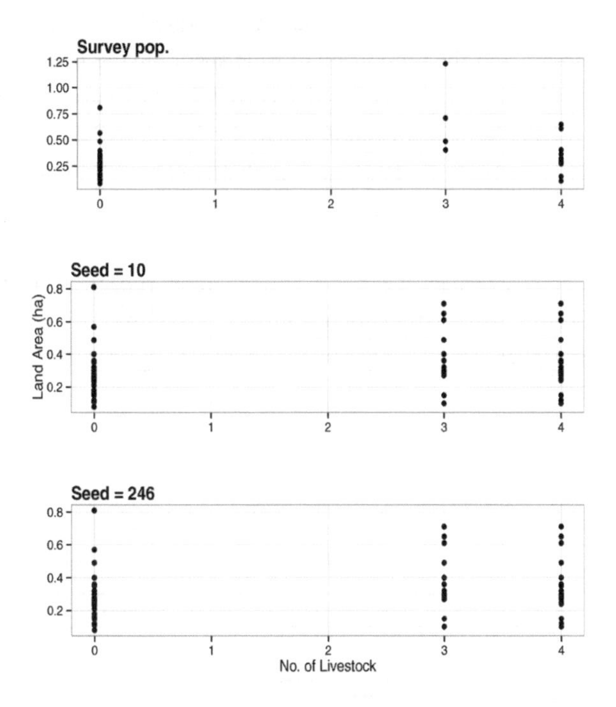

Figure 4 **Comparison of correlation between land area and number of livestock owned for both survey (A) and agent populations (B and C). Agent populations were generated using a random seed value of 10 and 246.**

Seasonal variation was also high for a number of the activities. This could be seen in the wealth rating (buy: sell ratio) of each agent type. Typically the buy: sell ratio was highest within the wet season, throughout the months of November to March, with an average and standard deviation of 5.1 (± 1.9), respectively. Following harvest of maize in the cold season however, the wealth rating was at its lowest for all agents, the average and standard deviation being 2.9 (± 0.84). Finally in the hot season, between the months of August to October all agent types experienced overall increases in buy: sell ratios, except those of type 1. Wealth ratings for types 2, 3 and 4 were between 35 and 92 per cent higher than the cold season, whilst the average buy: sell ratio for male headed households of medium and rich wellbeing decreased by 74 per cent.

The food-adequacy value on the other hand, demonstrated little seasonal variation. Instead, differences were apparent between the agent types (see Figure 5). Surprisingly, female headed households of rich and medium wellbeing reported higher food adequacy on average. Across 50 replicates and 10 years, agents of type 3 reported a mean and standard deviation of 1.17 (± 0.36); whilst corresponding male headed households (type 1) reported 1.04 (± 0.25).

A number of coping strategies emerged in response to drought stress. An increased dependence upon the market to buy maize was exhibited by all four agents over the ten years and across all 50 replicates. Linear regression analysis uncovered a significant relationship between drought probability and the proportion of time spent buying from the market ($F_{1, 70}$= 155.8, $p < 2e^{-16}$) in addition to a strong positive correlation with average agent wealth-rating ($F_{1, 70}$= 39.65, $p < 2.33e^{-8}$), reporting an R-squared value of 36.2 per cent.

One-way analyses of variance revealed timing of the input subsidy program had no significant effect on the proportion of time spent on maize cultivation, nor did it affect the average wealth rating of the four different agents (Dobbie, 2013). The proportion of agents who could access the input subsidies registered only a slight impact. A linear regression model revealed a relationship between the proportion of agents with access to input subsidies and time allocated to maize, which was only just significant ($F_{1, 70}$ = 4.2, $P < 0.04$). Increases

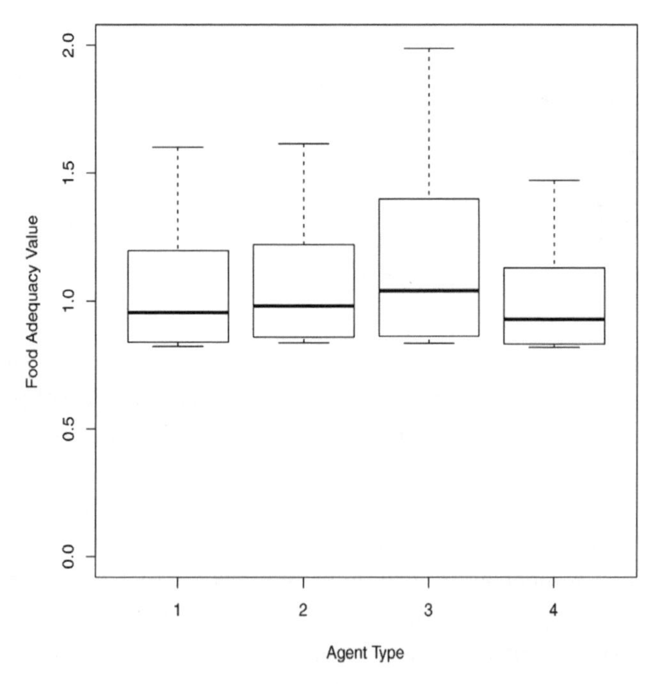

Figure 5 **Box Plot of Average Annual Food Adequacy Value according to Agent Type. Results are averages across 50 replicates of 10 years. Food adequacy value of 1 implies less than adequate supply of food; whereas a value of 2 implies supply of food is just adequate.**

in access to inputs did lead to reductions in time allocated to crops such as cassava (Dobbie, 2013). A negative correlation was also found between the average time spent cultivating cassava and the proportion of agents with access to inputs. Regarding seasons, in general, the impact of access to inputs was greatest in the wet season, between the months of November to March (Dobbie, 2013).

4. Discussion

A key aim of the study was to determine whether it is possible to employ PRA exercises in the construction of empirical agent-based models to explore complex social, ecological and political aspects of rural food security. Overall, results of this report show support for this novel application of PRA exercises, however, a number of issues have been raised. The remainder of this section will take a critical look at the methods used to construct the ABM from PRA exercise data. In particular, the assumptions underlying the steps required for agent type parameterisation and allocation will be examined. Additionally, attention will be given to the ABM's potential for investigating the impact of policy upon household decision-making and food security.

4.1. Characterisation of Empirical Agents

Design of the PRA exercise did prove to facilitate greater understanding of small holder decision making under stress and a number of interesting discoveries were made. The motivation behind crop and livelihood diversification for instance, was not to improve a household's self-sufficiency, but to propel additional maize purchases. Throughout both baseline and drought years, villagers planted crops such as sweet potato and cassava, collected wild or indigenous fruits and fetched additional fire wood in order to create tangible assets to be sold at market. The majority of income generated by these activities was then used in the purchase of additional maize stocks.

The cultural importance of maize was also highlighted by the responses of PRA participants to the interview-style questions. Despite knowledge of the drought tolerant qualities of crops such as sweet potato and cassava being widespread amongst farmers, when asked if they would switch cultivation in favour of such crops, the consensus was that they would prefer to continue with improved maize varieties. The reluctance of villagers within Southern Malawi to include cassava within their diets was also uncovered by ASSETS. When constructing a food security timeline as part of exercise P of ASSETS PRA manual, one participant from village 4 quoted "This cooked cassava also upsets the stomach because we are just used to consuming maize" (Schreckenberg *et al.*, 2012).

In terms of implementation, typically the PRA exercises took between 90 minutes and 2 hours to complete. This posed a time constraint to the field work and limited the ability to 'triangulate' results. Triangulation is a fundamental aspect of PRA techniques in which the validity and reliability of information is secured (Chambers, 1994). Cross-checking is typically achieved through repeated discussions of the topic with different participants and the use of complementary data collection methods, such as mapping, diagrams, matrix scoring etc. (Chambers, 1994). For this project, the scope of triangulation was reduced to simply repeating the PRA exercises with participants from different villages and consolidating information garnered from task one, with interview questions in task two (and *vice versa*). Results from this study however, identified significant inter-village differences (See Table 5). Triangulation conducted in this manner therefore, is not sufficient to validate PRA findings.

Results from the PRA exercises were used to create seasonal behavioural rules dictating the amount of effort given to productive activities (See Table 3). Here, the proportion of effort an agent spends on a particular household activity such as maize farming and selling at the market etc. is drawn from a log-normal distribution. The mean and standard deviation of the distribution is calculated from results corresponding to the activity and agent type in question across the four villages studied.

Underlying assumptions of the behavior rules include that overall effort is uniform all year round. For each month, the number of counters to be divided between the activities remained the same. Uniform distribution of effort throughout the year is likely to be an oversimplification as seasonality affects the productivity of households (Wodon & Beegle, 2006). Furthermore, the task focused upon time allocated to productive activities only. The importance of cultural activities was not encompassed by the exercise.

In addition, it is assumed that the allocation decisions of PRA participants belonging to the same gender and wellbeing grouping, but from different villages, fall within the same distribution. Results from the one-way analyses of variance for inter-village variability of each of the activities, reveal a caveat in the models construction (See Table 5). Highly significant results suggest that assuming the time allocation decisions of participants from different villages can be drawn from equivalent distributions is flawed. A number of ecological and social factors may lead to behavioural differences between villages. Such differences pose difficulties when attempting to scale agent types effectively.

In general, data scarcity provided a number of issues when generating agent populations at the Traditional Authority level. Female headed households for example, were poorly represented within survey data corresponding to the study area of interest. As a result, the agent population is less robust for female headed households of either medium or rich wellbeing and poor or very poor wellbeing (see Figure 3). The underrepresentation of marginalized groups may act to undermine the worth of empirical ABM for policy analysis.

Advances in empirical data collection techniques however, could help overcome data scarcity issues. A key advantage of agent-based modeling is the ability to integrate both qualitative and quantitative data from a large number of sources (Liu et al., 2008). This study demonstrates the use of both PRA data and household survey data. There is scope to use additional methods such as remote sensing and statistical analysis of spatial data (Brown *et al.*, 2008).

4.2. Analysis of Household Food Security

A further aim of the study was to investigate whether the model could be used to explore household food security. It is important to acknowledge that the ABM described throughout this project, represents a model in its early stages of development. Further verification and validation steps will be required before plausible conclusions and policy recommendations can be drawn from model outcomes.

Results of the three scenarios: baseline, drought and input subsidies do however permit some speculation regarding the potential of maize to assist smallholders in the realisation of food security. Increasing the probability of drought for example, revealed a negative correlation with the proportion of time spent on maize, as well as concurrent decreases in the food adequacy value of each of the agent types. Regarding political factors, reliance upon an input subsidy programme for fertilisers and improved maize varieties can be explored through results from the input subsidy scenario. A key finding here is that unless the proportion of smallholders with access to inputs was above 60 per cent, the programme made little impact on the average food adequacy value of agents. This highlights the unsustainable nature of the farm input subsidy scheme, as high operational cost are required in order to target households in sufficient number for it to be of any benefit to national food security.

As it stands, a simple food adequacy variable is not a sufficient measure of household food security. It fails to encompass availability, access, utilization and stability dimensions of food security sufficiently. Robust testing of the model also falls out of the scope of this report; implementation involved only a sensitivity analysis of model parameters. The impact of behavioural decisions upon model outcomes was not considered, neither were the interactions between different parameters. Instead, the focus here was on the way in which empirical agent based models should be constructed. Future work will therefore expand upon the food adequacy variable to provide a measure that reflects the multi-dimensional nature of household food security; as well as broach the subject of model validation, an issue which is hotly debated throughout the literature (Windrum *et al.*, 2007; Moss, 2008; Becu *et al.*, 2003). Validation of the model by both the villagers who assisted in its creation and other, potential end users would be in line with the underlying PRA ethos, whereby data is scrutinized, retained and shared by local people in a bid to propel empowerment.

5. Conclusions

A key aim of this study was to determine whether an empirical ABM of rural households could be created in a systematic manner and at a level relevant to policy analysis. Overall, results support the use of participatory approaches and demonstrate the value of recent frameworks and guidelines for the creation of empirical ABM. In general, methods advocated by Smajgl *et al.* (2011), Valbuena *et al.* (2008) and Berger & Schreinemachers (2006), did support the creation of a robust ABM. However, caution must be heeded when attempting to scale agent types. Significant inter-village differences can be overlooked when generating an agent population at the Traditional Authority level. This highlights a trade-off between creating an ABM at a level relevant to policy analysis while providing an accurate representation of system heterogeneity. Data scarcity may also hinder the representation of marginalized groups, particularly within developing country contexts (Berger & Schreinemachers, 2006). Continued advancement of data collection techniques does however hold potential to bridge this gap. In addition to PRA data, techniques such as remote sensing and statistical analysis of spatial data may be used for model calibration and validation (Brown *et al.*, 2007).

Once built, initial results of the model did permit inferences to be made concerning the impact of drought upon rural smallholder decision making in Southern Malawi. It also allowed the impact of the farm input subsidy program to be investigated. However, comprehensive model verification and validation was not within the scope of the project and so the results must be treated with caution.

In summary, the project highlights the potential of empirical agent-based modelling for the analysis of data from PRA exercises and household surveys. It provides a generic methodology which can be adapted and applied to other regions of interest. The use of participatory techniques to identify household types and decision rules may be of particular relevance within developing and / or data scarce environments. Attention now turns to the role of the technique in providing a collaborative learning framework which promotes interaction between scientists, policy makers and stakeholders alike. Under such a framework, it is hoped that integration of techniques within the social sciences, and tools from the realms of complexity science may finally assist rural smallholders in the realisation of the four pillars of food security.

Notes

1 http://espa-assets.org/
2 http://www.openabm.org/model/3946/version/1/view

References

An, L. 2012. "Modeling human decisions in coupled human and natural systems: Review of agent-based models". *Ecological Modelling*, *229*, 25–36. doi:10.1016/j.ecolmodel.2011.07.010

Becu, N., Perez, P., Walker, A., Barreteau, O., and Page, C. L. 2003. "Agent based simulation of a small catchment water management in northern Thailand". *Ecological Modelling*, 170, (2-3), 319–331. doi:10.1016/S0304-3800(03)00236-9

Berger, T., and Schreinemachers, P. 2006. "Creating agents and landscapes for multiagent systems from random samples". *Ecology and Society*, 11, (2), 19.

Berger, T., Schreinemachers, P., and Woelcke, J. 2006. "Multi-agent simulation for the targeting of development policies in less-favored areas". *Agricultural Systems*, 88, (1), 28–43. doi:10.1016/j.agsy.2005.06.002

Bohnet, I. C., Roberts, B., Harding, E., and Haug, K. J. 2011. "A typology of graziers to inform a more targeted approach for developing natural resource management policies and agricultural extension programs". *Land Use Policy*, 28, (3), 629–637. doi:10.1016/j.landusepol.2010.12.003

Breisinger, C., Ecker, O., Al-Riffai, P., and Yu, B. 2012. *Beyond the Arab awakening: policies and investments for poverty reduction and food security*. International Food Policy Research Institute. Washington DC

Brown, D. G., Robinson, D. T., An, L., Nassauer, J. I., Zellner, M., Rand, W. and Wang, Z. 2008. "Exurbia from the bottom-up: Confronting empirical challenges to characterizing a complex system". *Geoforum*, 39, (2), 805-818.

Chambers, R. 1994. "The origins and practice of participatory rural appraisal". *World development*, 22, (7), 953-969.

Chibwana, C., Fisher, M., and Shively, G. 2012. "Cropland Allocation Effects of Agricultural Input Subsidies in Malawi". *World Development*, *40*(1), 124–133. doi:10.1016/j.worlddev.2011.04.022

Deadman, P., Robinson, D., Moran, E., and Brondizio, E. 2004. "Colonist household decisionmaking and land-use change in the Amazon Rainforest: an agent-based simulation". *Environment and Planning B*, 31, 693-710.

Dobbie, S.L. 2013. *Simulating Household Decision Making in Rural Malawi; Empirical Characterisation of Agent Behaviour During Times of Drought Stress.* May 2013, available at http://www.openabm.org/model/3946/version/1/view

Dorward, A. R. 2003. *Modelling poor farm-household livelihoods in Malawi: lessons for pro-poor policy. Wye, Ashford, UK, Centre for Development and Poverty Reduction, Department of Agricultural Sciences, Imperial College London.*

Dorward, A., and E. Chirwa. 2012. "Evaluation of the 2010/11 Farm Input Subsidy Programme, Malawi: report on Programme Implementation." Paper prepared for MoAWID, Malawi and DFID. London, School of Oriental and African Studies.

EM-DAT (2013) *The OFDA/CRED International Disaster Database*, May 2013, available at www.emdat.be

Farmer, J.D. and Foley, D. 2009. "The economy needs agent-based modelling". *Science*, 460, 685–686.

FAO 1996. *Rome declaration on world food security and World Food Summit Plan of Action.* World Food Summit 13–17 November 1996. Food and Agriculture Organization, Rome, Italy.

FAO 2008. *State of food insecurity in the world. Rome: FAO.*

Grimm, V., Berger, U., Bastiansen, F., Eliassen, S., Ginot, V., Giske, J., ... and DeAngelis, D. L. 2006. "A standard protocol for describing individual-based and agent-based models". *Ecological modelling*, 198, (1), 115-126.

Grimm, V., Berger, U., DeAngelis, D. L., Polhill, J. G., Giske, J., and Railsback, S. F. 2010. "The ODD protocol: a review and first update". *Ecological Modelling*, 221, (23), 2760-2768.

Haddad, N., Duwayri, M., Oweis, T., Bishaw, Z., Rischkowsky, B., Hassan, A. A., and Grando, S. 2011. "The potential of small-scale rainfed agriculture to strengthen food security in Arab countries". *Food Security*, *3*(1), 163-173.

Holtz, G., and Pahl-Wostl, C. 2012. "An agent-based model of groundwater over-exploitation in the Upper Guadiana, Spain". *Regional Environmental Change*, 12, (1), 95-121.

Ibnouf, F. O. 2011. "Challenges and possibilities for achieving household food security in the Western Sudan region: the role of female farmers". *Food Security*, *3*(2), 215-231.

Janssen, M. A., and Ostrom, E. 2006. "Empirically based, agent-based models". *Ecology and Society*, 11, (2), 37.

Johnson, M. E., Masters, W. A., and Preckel, P. V. 2006. "Diffusion and spillover of new technology: a heterogeneous-agent model for cassava in West Africa". *Agricultural economics*, 35, (2), 119-129.

Le, Q. B., Seidl, R., and Scholz, R. W. 2012. "Feedback loops and types of adaptation in the modelling of land-use decisions in an agent-based simulation". *Environmental Modelling & Software*, 27, 83-96.

Liu, Y., Gupta, H., Springer, E., and Wagener, T. 2008. "Linking science with environmental decision making: Experiences from an integrated modeling approach to supporting sustainable water resources management". *Environmental Modelling & Software*, 23, (7), 846-858.

Masters, W., Shively, G., Hogset, H., and Fisher, M. 2000. *Evaluating Recent Strategies to Improve Food Security Among Smallholder Households In Malawi.* USAID–IFPRI

Mena, C. F., Walsh, S. J., Frizzelle, B. G., Xiaozheng, Y., and Malanson, G. P. 2011. "Land use change on household farms in the Ecuadorian Amazon: Design and implementation of an agent-based model". *Applied Geography*, 31, (1), 210-222.

Moss, S. 2008. "Alternative approaches to the empirical validation of agent-based models". *Journal of Artificial Societies and Social Simulation*, 11, (1), 5.

NSO 2012. *Republic of Malawi, Integrated Household Survey 2010 -11.* August 2013, available at http://www.nsomalawi.mw/index.php/third-integrated-household-survey-ihs3.html

Njaya, F. 2007. "Governance Challenges of the Implementation of Fisheries Co-Management: Experiences from Malawi". *International Journal of the Commons, 1*(1), 137-153.

Parker, D. C., Manson, S. M., Janssen, M. A., Hoffmann, M. J., and Deadman, P. 2003. "Multi-agent systems for the simulation of land-use and land-cover change: a review". *Annals of the association of American Geographers*, 93, (2), 314-337.

Robinson, D. T., Brown, D. G., Parker, D. C., Schreinemachers, P., Janssen, M. A., Huigen, M., ... Barnaud, C. 2007. "Comparison of empirical methods for building agent-based models in land use science". *Journal of Land Use Science*, 2, (1), 31–55. doi:10.1080/17474230701201349

Sahley, C., Groelsema, B., Assistance, H., Marchione, T., and Nelson, D. 2005. *The Governance Dimensions of Food Security in Malawi*, USAID.

Schlüter, M., and Pahl-Wostl, C. 2007. "Mechanisms of resilience in common-pool resource management systems: an agent-based model of water use in a river basin". *Ecology and Society*, 12, (2), 4

Schreckenberg, K., Torres Vitolas, C.A., Willcock, S., Shackleton, C. and Harvey, C. 2012. *ASSETS field manual for community level data collection*, May 2013, available at http://gtr.rcuk.ac.uk/publication/00410038-0036-0030-4600-350035003700/

Smajgl, A., Brown, D. G., Valbuena, D., and Huigen, M. G. A. 2011. "Empirical characterisation of agent behaviours in socio-ecological systems". *Environmental Modelling & Software*, 26, (7), 837–844. doi:10.1016/j.envsoft.2011.02.011

Thangata, P. H., Hildebrand, P. E., and Gladwin, C. H. 2002. "Modeling agroforestry adoption and household decision making in Malawi". *Afr Studies Q. The Online Journal for African Studies. University of Florida.*

Valbuena, D., Verburg, P. H., and Bregt, A. K. 2008. „A method to define a typology for agent-based analysis in regional land-use research". *Agriculture, Ecosystems & Environment*, 128, (1-2), 27–36. doi:10.1016/j.agee.2008.04.015

Wodon, Q., and Beegle, K. 2006. *Labor shortages despite underemployment? Seasonality in time use in Malawi*, May 2013, available at http://mpra.ub.uni-muenchen.de/11083/

Windrum, P., Fagiolo, G., and Moneta, A. 2007. "Empirical validation of agent-based models: Alternatives and prospects". *Journal of Artificial Societies and Social Simulation*, 10, (2), 8.

Ziervogel, G., Nyong, A., Osman, B., Conde, C., Cortés, S. and Downing, T. 2008. *Climate Variability and Change: Implications for Household Food Security*, August 2013, available at http://www.aiaccproject.org/working_papers/Working%20Papers/AIACC_WP_20_Ziervogel.pdf

Appendices

1. Model Procedures

The procedures, *define-fields ()* and *define-agents ()* act to create the agent population and the environment in which they reside. The procedure *set-month (),* labels each time-step with its corresponding month – step 1 is January, step 2 is February etc. *Set-drought ()* on the other hand, acts to determine whether or not the current year is a drought year. If a randomly generated number is greater than the selected drought probability, drought is deemed to be true (and *vice versa*). The influence of input subsidies upon agents' decision making is established by the procedures *set-exo-impact ()* and *set-exo-onset ().set-exo-impact ()* employs a randomly generated number between 0 and 1. In a similar vein to *set-drought ()*; if the probability of access is less than or equal to the random number, access to inputs is deemed true (and *vice versa*). If an agent has access to inputs, the proportion of time spent farming maize and pigeon pea is increased by a factor labelled *EXO-M* and *EXO-P*, respectively (See Appendix 7.2. for values). Once the impact of the input subsidy programme has been registered, the procedure *set-agent-type-decisions ()* calculates the proportion of time spent on each of the activities by the agent. This is achieved by first totaling all of the time-allocation values for the activities and then using this to calculate the relative contribution of each activity. A further calculation is made by the procedure *calculate-agent-wealth* (), which determines the ratio of the two activities buy at market: sell at market for each agent. Finally, the procedure *calculate-agent-food-adequacy ()* determines the perceived food adequacy of the agent, where 1 represents less than adequate, 2 just adequate and 3, more than adequate. This is a simple procedure, which in a good year, when drought is false, input access true and wealth-rating <= 1; food adequacy will increase by 0.2; whilst in a bad year (when the opposite is true), food adequacy will decrease by 0.2.

```
Model Procedures

define-landscape ()
define-agents ()
LOOP
    set-month ()
    set-drought ()
    set-agent-type-options ()
    set- exo-onset ()
    set- exo-impact ()
    set-agent-type-decisions ()
    calculate-agent-wealth ()
    calculate- agent -food-adequacy ()
END LOOP
```

2. Model Parameters

Parameter	Description	Value	Source
P-DROUGHT	Likelihood of it being a drought year	0.3	EMdat (2013)
P-EXO	Number of households with access to inputs	0.47	Dorward & Chirwa (2012)
EXO-FACTOR-M	Access to inputs leads to an increase in time spent on maize by a factor	1.16	Chibwana *et al.* (2012)
EXO-FACTOR-P	Access to inputs leads to an increase in time spent on pigeon pea by a factor	1.05	Dorward & Chirwa (2012)

3. Model Sensitivity Analyses

Three key scenarios were explored to test sensitivity of model parameters.

Scenario	Description
Baseline	There was a 30 per cent chance of drought, input subsidies became available from September to December of each year and could be accessed by 47 percent of the agent population (Dorward & Chirwa, 2012). Acquisition of input subsidies led agents to increase time spent upon maize by 16 per cent (Chibwana *et al.*, 2012) and pigeon pea by 5 per cent (Dorward & Chirwa, 2012). After each time step, once time-allocation decisions had been made for each activity, a new wealth-rating and food adequacy value were calculated for each agent. The model was run for 120 time steps, corresponding to 120 months, or 10 years. A total of 50 replicates were simulated, using a different random seed value each time. Random seed variation tests for the effect of random elements in the model. Random variables include: random-drought employed in the procedure set-drought and exo-random, used to select a group of agents to be given input subsidies. For each replicate, at each time step the mean and standard deviation of: time spent upon each activity, wealth-rating and food-adequacy for each agent type was calculated.
Drought	The model was run in the same manner as the baseline scenario, except this time, exploring the impact of drought upon agent decision making. The probability of drought was varied from 0 to 100 per cent in 20% intervals; bringing the total number of runs to 300.
Input subsidy	The model was run again, using the same setup as the baseline scenario but this time investigating the effect of input subsidies. Three different scenarios concerning the timing of the input subsidy program first simulated, namely 'early', 'typical' and 'late' were employed. After this, agent access to input subsidies was varied from 0 to 100 per cent in 10% intervals. Finally, by simultaneously increasing time spent upon maize and pigeon pea between 0 and 100 per cent, again in 10 % intervals, the bias towards maize and pigeon pea under input subsidies was investigated.

11

Gulf Cooperation Council Initiatives: Towards a Coherent Integrated Plan for Utilising Renewable Solar Energy

Mayami Abdulla

1. Introduction

The Cooperation Council for the Arab States of the Gulf, also known as the Gulf Cooperation Council (GCC), was established in 1981. It comprises Bahrain, Kuwait, Oman, Qatar, the Kingdom of Saudi Arabia and the United Arab Emirates (UAE) – which is a federation that is comprised of seven emirates: Abu Dhabi, Sharjah, Ras Al Khaima, Fujairah, Umm Al Quwain, Ajman and Dubai. The Arab states of the Gulf are characterised as having substantial fossil fuel resources and their principal source of energy generation is derived from fossil fuels. Energy is a crucial commodity and we consume it in all aspects of our lives, from small domestic appliances and technologies such as electric toothbrushes and mobile phones to huge industrial plants. In general, the GCC countries have encountered unprecedentedly rapid socio-economic growth accompanied by energy consumption patterns that have created a surge in demand (Bachellerie, 2012). Furthermore, these countries vary in the time projected to reach a post-oil era and some possess only small endowments, which are expected to deplete within the next twenty years (Krane, 2012). Therefore, in view of worldwide concerns over energy supply, the GCC countries in parallel with other countries are working on reducing their reliance on fossil fuel. In this pursuit, renewable energy resources and principally solar energy stand out as an appropriate response since, on the one side, solar energy is a free, neat and non-diminishing alternative and, on the other side, the Gulf region receives an abundant amount of daily solar radiation. Furthermore, utilising solar radiation for power generation will assist the GCC countries to meet the challenge of continually increasing demand along with achieving the objective of social, economic and environmental pillars of sustainability (*The global partnership for environment and development: a guide to Agenda 21*, 1992).

However, despite the several initiatives that are proving useful in the GCC countries to employ renewable energy (mainly solar energy) to generate electricity, these countries still lack an integrated adoption plan that will enhance the domestic diffusion of solar appliances that would complement these initiatives. Therefore, the chapter inspects the energy policy in these states, indicating the measures that would help in the domestic uptake of solar energy appliances and consequently assist these states to accomplish the target goals of energy in line with their future economic vision.

The remainder of this chapter consists of the following sections. The second section discusses GCC countries' drivers to utilise renewable energy resources in the energy generation portfolio. The third section explains the advantages that the GCC countries are attaining from utilising renewable energy to generate electricity. The fourth section reports some of the forms of renewable energies that these countries have incorporated into the mix of resources. Some of the GCC countries' initiatives towards the use of renewable solar energy for electricity generation are conveyed in the fifth section. Their renewable energy policies are scrutinised in the sixth section. The chapter concludes with recommendations to improve energy policy measures towards renewable solar energy in the GCC countries.

2. The Drivers behind the GCCs' Utilisation of Renewable Resources in the Energy Generation Portfolio

The GCC countries' electricity generation hinges primarily on fossil fuel derivatives and typically on natural gas. The escalation of energy consumption has led to an increase of hydrocarbon fuel usage in response to mounting demand. In addition, the scarcity of water as a precious item for living is also a forcing parameter. Indeed, energy and water security are current pressing issues around the world in general and in the GCC countries in particular. Energy and water security are, in the GCC countries, bound up in convoluted and firm links. These countries lie in an arid region which, on the one hand, suffers from limited access to fresh water resources and low rain levels. On the other hand, the region as a whole faces an increase in salinity of the available fresh ground water (Nouh, 2008). To overcome this crucial and massive need, the GCC countries have established water desalination plants; with 33 desalination plants, Saudi Arabia is the world's leading producer of desalinated water (Taleb and Sharples, 2011). However, these plants, which are usually coupled with an electrical power generation plant, are fired by fossil fuel derivatives, with natural gas being the main feedstock. The costs of construction followed by the need for regular maintenances are high and the plants operate by intensive fuel intake. Hence, more strains are exerted on the supply of energy meaning that the unsustainability of the fossil-fed energy resources threatens the energy supply. These desalination plants are also found to intensify the salinity of seawater in the Gulf, which not only harms the marine environment, but might also have negative consequences on the functional efficiency of the plants (Dawoud and Mulla,

2012, McLauchlan and Mehrubeoglu, 2010, Nouh, 2008). Similar to domestic power, desalinated water is extensively subsidised and is provided at low cost relative to production cost, which has resulted in domestic overconsumption patterns. Also, the GCC countries have responded to increasing demand for water by building more desalination plants; hence, exerting more strain on the fossil fuel energy resources.

Moreover, in accordance with the target to diversify the economy, the GCC countries have succeeded in attracting energy-intensive industries. These industries have prospered from the minimal taxation policy and the reduced price of electricity due to the energy subsidies, which leads to intensifying the demands for energy (Bhutto *et al.*, 2014, Dargin, 2010, Fattouh and El-Katiri, 2013, Reiche, 2010a). Nevertheless, the shortage of fossil fuel which is crucial to operate these industries as well as the envisioned carbon tax to protect the environment and to mitigate climate change would be seen as a significant economical threat to some of these industries.

Additionally, domestic consumption accounts for the dominant electrical demand due to the hot climate conditions and is encouraged by the low cost of electricity. The bulk of household electrical energy is allocated for air-conditioning, as indicated by the rise of consumption during heat waves (Bachellerie, 2012, Krane, 2012). This has attracted attention to houses and the built environment and led to a call for sustainable architecture (Taleb and Sharples, 2011). For example, a non-compulsory thermal insulation code, recently introduced to building regulations in most of the GCC countries, was deemed to have curbed electricity consumption by 40% in Bahrain and the UAE (Sharples and Radhi, 2013). Yet, the current domestic consumption of electricity remains so high that some of the GCC countries have encountered or are at risk of rolling blackouts during long hot summers (Bachellerie, 2012, *Bahrain boils in summer power cut*, 2004, *Kuwait may face electricity blackouts* 2013).

Furthermore, the lack of reliable public transport is regarded as having an indirect negative influence on energy consumption. Commuters are entirely dependent on domestic transportation and so there is a high demand for fossil fuel, especially in the geographically larger countries such as Kuwait and Saudi Arabia. Such high consumption of fossil fuels for transport is extensively subsidised and so adds fiscal impediments to national budgets (Krane, 2012, Lahn and Stevens, 2011, Lahn *et al.*, 2013).

In the past, the GCC countries enjoyed domestic self-sufficiency in natural gas. However, in the present situation the huge demand for natural gas as a feedstock for energy generation has surpassed the available low-cost domestic supply of natural gas in some of them. Confronted by shortages of natural gas resources, some of these countries have shifted from exporting to importing natural gas at the market prices (Bachellerie, 2012, Krane, 2012).

All the consumption channels of energy discussed above are placing unprecedented pressure on the GCC countries; national budgets, which **are** substantially contingent on oil, are under stress due to the massive subsidies given to fossil fuel energy, which are, in turn, placing a huge burden on the budgets. Moreover, the influence of the prevailing international circumstances of volatile fluctuation in fossil fuel prices is, needless to say, correspondingly impactful and adding a weight to the persistent contemporary pressures.

3. The Benefits for GCC Countries of Utilising Renewable Energy

By 2006, all GCC countries had ratified the Kyoto protocol and, by 1996, they had all agreed to the United Nations Framework Convention on Climate Change (UNFCCC), which obliges them to mitigate their greenhouse emissions. Therefore, in reaction to all their insistent burdens, besides their commitment to these international agreements, the GCC countries have explored other energy resources as substitutes for hydrocarbon fuels to remove the sole dependency on fossil fuels and to pave the path towards mixed fuel energy resources, also referred to as an "energy portfolio".

The fossil fuels used for electrical power generation account for ~42% of global carbon dioxide (CO_2), which adds to the greenhouse effect and leads to global warming (*CO2 Emissions from fuel combustion - highlights*, 2013). Furthermore, the emission of CO_2 from the GCC countries is among the highest levels globally (Sharples and Radhi, 2013). Adoption of renewable energy including solar energy for power production on a utility-scale and at the domestic level will certainly have positive environmental impacts by reducing CO_2 emissions. For example, it is anticipated that the utilisation of renewables into the energy portfolio in the United States will help to cut CO_2 emissions in 2050 by 62% of the level in 2005 (Fthenakis, 2009, Fthenakis *et al.*, 2009).

The use of renewable energy resources will help to extend the longevity of the contemporary available fossil endowments through preserving them. Furthermore, it will permit the GCC monarchies to allocate more of the produced fossil fuel towards exports. This would be economically preferable to using the fuel to feed electricity generation plants since the actual cost of the generated electricity cannot be redeemed owing to the heavy energy subsidies (Kemp, 2014). Therefore, a more diversified energy portfolio will certainly have a direct impact on reinforcing the economy of these states by reducing the high cost of fuel subsidies and will help them to sustain their economic leadership in the world. In addition, renewable energy will prepare the GCC states for the post-oil era when all fossil fuels have been depleted. For example, it will aid in reorienting the available wealth and revenue towards investment in infrastructure as well preparing the groundwork for a gradual, graceful transition and integration of solar and other energy sources into the energy portfolio.

4. Examples of the GCC Countries' Measures Towards a Mixture of Energy Resources

In parallel to the international trend on the route to establishing an energy portfolio, the GCC countries started several initiatives in tapping into various energy resources and some are currently working on expanding their present projects. Resources that are alternative or complementary to fossil fuel that have been considered by these countries are nuclear energy (non-renewable) as well as wind and solar energy (renewable). A few examples of their attempts and initiatives are briefly discussed here with an emphasis on solar energy.

The recent interest of GCC countries in nuclear energy can be traced to the meeting of GCC leaders in 2006, when the intention of launching a nuclear energy development programme was declared. However, the Gulf countries diverge when it comes to adopting the nuclear power option (Emerging Nuclear Energy Countries, 2014). The UAE is the most proactive, taking the lead with the establishment of its first power station in July 2012, which is planned to be operational in 2017. This plant was followed by the building of the second in May 2013 and third in 2014, with a stated goal of having four plants by 2020 (Emerging Nuclear Energy Countries, 2014, *UAE nuclear plant unit 3 work starts*, 2014). In 2010, Bahrain announced the adoption of nuclear energy for power generation by 2017; however, the plan was postponed in 2012 (Omari, 2012). Likewise, Kuwait and Oman reversed their decisions to incorporate nuclear energy into the proposed future vision of mixed energy resources after the accident at the Fukushima nuclear plant in Japan (Ahmad and Ramana, 2014, Krane, 2012). Saudi Arabia has performed studies for the feasibility of nuclear energy and its envisioned contribution towards the non-fossil future energy portfolio in 2032. It has been reported that a survey to evaluate and select plant sites is to be prepared for the construction of the first nuclear energy plant. Although it is not clear whether this proposal has been approved, some analysts recommend that the renewable energy sources are more favoured in the short term than the nuclear energy option (Ahmad and Ramana, 2014, Krane, 2012). However, other sources report that the construction of the nuclear plant will start between 2016 and 2017 with the aim of completion by 2022 (*Nuclear power in Saudi Arabia*, 2014, Yee, 2013).

The political sensitivity of the area, however, makes nuclear energy in the GCC states debatable. In addition, the huge initial cost for the construction of nuclear energy plants and all the associated expenditure for maintenance along with the set up and implementation of mandatory safety measures constitute parameters that compel the GCC countries, except the UAE, to dampen or postpone their interest in nuclear and to divert their attention towards renewable energy.

Renewable wind energy is found to be a moderate source of power in Bahrain and is currently employed in the Bahrain World Trade Center to generate 15% of the Center's total electricity consumption (*Bahrain World Trade Center*). Studies of wind energy in Kuwait, Saudi Arabia, Oman and Qatar demonstrate that wind energy is economically feasible. Also, studies have shown that some parts of the UAE have economically viable potential wind energy, e.g., where a wind turbine was installed on Sir Bani Yas Island (Bachellerie, 2012, Bhutto *et al.*, 2014).

Solar energy stands out as the most applicable alternative and best source of environmentally friendly energy for the GCC countries. Indeed, in pursuing a practical and low-cost energy resource it would be best for the Gulf monarchies to exploit the abundant amount of sunlight with which they are blessed, combined with the availability of spacious lands (for solar farms) in some of these countries (Whittell, 2014). In addition, the heightened domestic demand for electricity in the GCC countries coincides with the summer, which is characteristically hot. Moreover, the sky is clear, so these countries would benefit from a total amount of solar radiation unrivalled in the world. Solar radiation is harvested using these methods and appliances: photovoltaics (PVs), concentrator solar power (CSP) systems and solar water heating (SWH). PVs, for example, are safe and exploit solar energy, which is freely available. Moreover, PVs require only a small cost for maintenance and have an estimated life expectancy of 20-30 years, which may make them economically viable in the long term (Mekhilef *et al.*, 2012). Furthermore, the rich solar resources in Gulf countries give them the potential to be vendors of solar energy by exporting a share from the energy that would be generated utilising solar appliances. Therefore, in parallel with the future aim of the "The Meditation Solar Plan" project to export part of the generated solar energy, some investors in the GCC region discussed the idea of trading solar energy. Finally, the potential of exporting solar energy was also highlighted by Khalid Al Sulaiman, vice president for renewable energy at the King Abdullah City for Atomic and Renewable Energy (KA.CARE) in Saudi Arabia (Clercq, 2013, Jablonski *et al.*, 2012, Reiche, 2010a).

5. Examples of the GCC Countries' Initiatives Towards Renewable Solar Energy

The steps of the GCC countries in renewable solar energy initiatives align well with the goals of green economy and sustainable developments (*The global partnership for environment and development: a guide to Agenda 21*, 1992, *Green economy report: a preview*, 2010). In 2009, Abu Dhabi succeeded in hosting the headquarters for the International Renewable Energy Agency (IRENA) at Masdar City, in a step that will support the renewable energy endeavours in the region. The studies and initiatives pursued by Bahrain to utilise solar energy resources will be presented here briefly together with some very recent endeavours of the other Gulf states. For complete detailed information on the projects pursued by each of the GCC countries, the reader is advised to refer to the literature (Bachellerie, 2012, Bhutto *et al.*, 2014, Bollier, 2014, Hepbasli and Alsuhaibani, 2011, Patlitzianas *et al.*, 2006, Reiche, 2010a).

Among the GCC countries, Bahrain took the lead in exploring the viability of solar power through establishing a gas station powered by PVs in early 1980, which operated for a few years and was then paused (Al-Jayousi, 2012). Also, several studies related to the feasibility and exploitation of renewable energy were conducted by staff at the University of Bahrain and through a foreign consulting firm (Al-Qahtani, 1996, Alnaser, 1995, Alnaser and Aldudiafa, 1990, *H.E. The Ministry of Energy signs a Contract for Conducting Feasibility Study for Renewable Energies and Supervision of Establishment of a Pilot Power Plant*, 2011).

According to Bahrain's Economic Vision 2030, the country plans to produce 5% to 10% of its total electrical requirement utilising renewable energy sources, of which some will be used to produce desalinated water (Bhutto *et al.*, 2014, Mayton, 2012, Woods, 2014). In recent years, two solar energy projects have been proposed: one in 2012 in Awali city to produce five megawatts of energy, which is capable of supplying the whole energy demand for the residents of the city; it became operational in June 2014 (Al-Jayousi, 2012, Woods, 2014). The other project was planned in 2013 for the Hawar Islands and at the time of writing, a feasibility study for the project is being conducted by Masdar, a company owned by UAE (Al-Jayousi, 2013, Rafique, 2012). Furthermore, Bahrain's Ministry of Energy proposed conducting a study to investigate the attainability of solar-powered street lighting (*Major boost for renewable energy in Bahrain*, 2014, *Solar-powered street lighting explored* 2014).

Qatar's economic vision aims to utilise solar energy to generate 16% of its total electricity production by 2018 (*OPEC member Qatar says to start embracing solar energy*, 2012). Recently, the electricity and water company in Qatar declared a major aspirational solar energy plan involving the exploitation of the roofs of several of the state's existing utility water reservoirs to mount PVs (Clover, 2013). The target of Kuwait's energy plan is to produce 1% of the total energy using solar and wind renewable resources by 2016 with the stated aim to raise this contribution to 15% by 2030 (Meza, 2013). In order to achieve the targeted share of renewable energy, Kuwait has recently concluded the installation of its first solar power plant (*Kuwait's first PV plant underpinned by skytron energy's monitoring solution*, 2013). The Sultanate of Oman aims to produce 10% of its total future energy demand from renewable resources and launched, in 2013, a pilot project to utilise renewable solar and wind energy that is anticipated to provide Al-Mazyunah town with most of its power demands (Bachellerie, 2012, Riyami, 2013).

The total solar power capacity and investment of the UAE via its several renewable energy projects, primarily in Abu Dhabi and Dubai, which are either under implementation or in operation, have positioned it as third in the world in solar energy after the US and Spain (*UAE ranks third in the world for total solar power capacity and investment*, 2014). In the UAE, Abu Dhabi aims to reach a 7% energy production from renewable energy by 2020 (Saadi, 2014) whilst Dubai's goal is to produce 5% of its energy from renewable resources by 2030, to which end it has recently released a tender for constructing the Mohammad bin Rashid Al Maktoum Solar Park project (Clover, 2014b, Willis, 2012). Additionally, the UAE through Masdar City has launched one of the world's largest solar power stations, Shams-1, to generate electricity, which will be followed by Shams-2 and Shams-3 (*Shams solar power station*).

Saudi Arabia is working proactively to meets its 2032 target of obtaining a total of 54 gigawatts from renewable resources, of which 41 gigawatts would be produced from solar energy employing both PVs and CSP (Clercq, 2013). An example of a project to source electricity from PVs has been accomplished by a Shell subsidiary which has successfully employed rooftop PVs. This Shell subsidiary utilised the shaded car park array which extends over 16 hectares to implement PVs that fruitfully generate 100% of the Al-Midra office complex's daytime electricity demand (Woods, 2013). Another example is the 3.5 megawatt solar energy initiatives in Riyadh, which is the largest project for ground-mounted solar panels in Saudi Arabia (Stuart, 2011). Moreover, Saudi Arabia is also planning to exploit solar energy in water desalination, exampled by the plan to construct the world's largest desalination plant in Al-Khafji (*Largest solar powered desalination plant to be built in Al-Khafji*, 2013).

6. Renewable Solar Energy Policies in the GCC Countries

Energy policy measures are the strategies and tools that permit countries to tackle energy problems along the path from production to consumption and will endorse the efficient utilisation of resources as well as setting targets (Bhutto *et al.*, 2014). Therefore, energy policy measures, including those for solar energy, are comprised of energy generation vision and portfolio, policy legalisation, incentives for investment, guidelines for energy conservation, etc. As domestic electricity consumption constitutes an immense amount of the total energy consumption in the Gulf region, the discussion of solar energy policies will be limited to those relevant to the domestic level.

Despite examples of GCC states' initiatives in exploiting renewable energy in the form of small pilot projects or utility-scale ventures, the energy policy and the legislation provisions that would endorse an extensive deployment of renewable energy at a domestic level are very low (Patlitzianas *et al.*, 2006). According to policy transfer theory, the introduction of relevant policies will eventually lead to a wider placement via adoption by neighbouring countries (Fattouh and El-Katiri, 2013, Reiche, 2010a, Reiche, 2010b). Such transfer is signified in GCC countries by the contemporary transmission and adoption of renewable energy initiatives. Therefore, it is anticipated that the introduction of energy policy measures that enhance the domestic adoption of renewable solar energy in one country will be followed by rapid propagation to other countries in the region.

At the moment, the domestic adoption of renewable energy is very low; it barely exists in the GCC states and lacks the lavish subsidies that favour conventional fossil fuels. The conventional fossil fuel electricity is heavily subsidised in these states where the actual cost of power production is not being redeemed. Hence, these subsidies constitute a heavy burden on the national budgets of the GCC countries. This is because, as electricity increases annually, the subsidies progressively deplete the revenues of GCC monarchies. Moreover, energy subsidies adversely affect the economy as a result of encouraging energy overconsumption patterns that may jeopardise the amount of fossil fuel for export, and shift some states from net exporters to net importers for energy fuel (Bachellerie, 2012, Krane, 2012, Lahn and Stevens, 2011). Also, in view of assumed escalation in future energy demand, the sustainability of the subsidies towards fossil fuel energy is certainly becoming questionable from a budgetary aspect, especially considering that GCC countries' budgets are chiefly sourced by fossil fuel revenue and these countries vary in the time projected to reach a post-oil era (Bachellerie, 2012, Krane, 2012). In spite of calls for GCC states to reform and possibly revoke some of the heavy subsidies of fossil fuel, the countries are currently finding it difficult to amend these subsidies, both economically and politically (Dargin, 2010, Fattouh and El-Katiri, 2013, Kemp, 2014, Law, 2012, Wynn, 2014). Therefore, GCC countries should capitalise on some of the presently accessible revenues from hydrocarbon fuels in embarking on a policy that represents a coherent, integrated comprehensive plan for endeavours towards the usage of solar energy. These policy instrumentations will permit the GCC countries to attain gradual preparation for the post-oil era in terms of infrastructure, technical expertise and investment in research and development (R&D) as well as in their youth capital to supply the necessary skilled professionals. Therefore, keeping pace with the export capacities and subsequent revenues are compelling drivers for the instalment of measures supporting the wide deployment of solar appliances.

A study was performed in Bahrain to investigate the constraints that hinder the dissemination of solar and wind appliances in built environments from the perspectives of policy- and decision-makers, consultants and contractors. The policy- and decision-makers found that the lack of expertise and the high initial cost of these appliances hampered their uptake whilst the architects reasoned that this was down to the absence of sustainable design knowledge and energy economy. The contractors suggested that the lack of skilled professional could be resolved by either sub-contracting or introducing subsidised training programmes for the workforce (Alnaser and Flanagan, 2007).

Globally, the initial cost of implementing solar appliances is the main impediment to broad domestic endorsement. Therefore, several countries have established supporting policies in order to pave the path and foster the domestic deployment of solar energy by making it economically attractive to individuals. For GCC countries, the introduction of renewable energy support schemes is expected to facilitate a gradual reform of the existing fossil fuel subsidies, e.g., constraining their use to citizens with a low income would be a much smoother and publically acceptable approach. This might also lead to a steady transition of allocated subsidies budget from subsidising fossil fuel towards subsidising solar energy. Examples of policies that have been implanted around the world and assist in nurturing the uptake of solar energy domestically are financial subsidy programmes that aim to make solar energy economically viable through deducting the cost of solar energy utilising: renewable portfolio standards (RPS), subsidies, incentive tax credits, feed-in tariffs (FITs), etc. (Bhutto *et al.*, 2014, Mekhilef *et al.*, 2012, Solangi *et al.*, 2011). The distribution of photovoltaics (PVs) around the world is primarily derived by the FITs policy which is to be adopted in a total of 45 countries including Spain, Germany and Malaysia (Bhutto *et al.*, 2014, Solangi *et al.*, 2011). The FITs is a very strong policy measure whereby the owner of a renewable energy system such as PVs is involved in a long-term contract by which the service or regulated body pays the owner a premium price for the electricity generated (Mekhilef *et al.*, 2012).

The renewable portfolio standard (RPS) is a policy measure to incentivise the incorporation of renewable energy within the accumulative energy production and is exploited in, for example, the United States and United Kingdom (Solangi *et al.*, 2011, Zhai, 2013). The RPS are rules set by the regulatory authorities and they place the electricity supply companies under an obligation to generate a quantified fragment of electrical power utilising the various renewable energy resources such as solar, wind, etc. To illustrate, the owner of the energy appliances is provided with certificates based on the units of electricity being produced in their premises, where they can trade the gained certificates together with the produced electricity to the supply companies. Then these companies forward the certificates to a regulatory body to demonstrate their fulfilment of the regulatory requirements. Therefore, the issued certificates constitute an incentive for individuals to invest in PV owing and the opportunity to trade the certificates with the energy supply utility in order to comply with regulatory obligations (Swift, 2013).

A study in Taiwan has shown that the continuation of the government incentive programme in the form of a subsidy policy allocated to the purchasing and installation of solar water heaters will help the government to achieve its aim of an aggregate installation area of solar heaters by 2020 (Chang *et al.*, 2013). Another study, conducted in the United States, shows that fast placement was demonstrated in the states that utilised financial cash incentives in terms of rebates and grants as a supporting policy for solar photovoltaic diffusion (Sarzynski *et al.*, 2012).

Other policy measures are incentives such as tax incentives and investment grant to promote commercial diffusion of solar energy appliances by investors, such as used in Canada and Germany. In addition, governments provide subsidies that are allocated for Research and Development (R&D) of the PV technologies (Solangi *et al.*, 2011). In response to these two policy measures, one finds that some GCC countries have already taken actual steps to accomplish them. For instance, Qatar announced the commencement of a PVs factory and Saudi Arabia has recently announced an agreement with a firm from the United States to conduct a feasibility study into constructing a PVs manufacturing plant (Clover, 2014a, Meza, 2014). A study in Canada has shown that adopting policy measures that enhance PV manufacturing has economic as well as social and environmental benefits (Solangi *et al.*, 2011). Therefore, GCC countries are expected to gain similar benefits, as such a step, for example, will revoke the cost of importing PVs and will create jobs. With regard to R&D, the countries have realised the importance of collaborating with leading universities and industrial firms in the field of solar power in order to attain the latest cutting-edge knowledge and skills. Presently, several research funding bodies and agencies in the GCC countries consider renewable solar energy as a national priority, and it is topping their research agenda. Therefore, huge investments from funding agencies are dedicated towards R&D carried out in universities and institutions in the region, some of which are collaborative efforts with leading international counterparts (further detailed information is available in the literature:(Bachellerie, 2012, Bhutto *et al.*, 2014, Patlitzianas *et al.*, 2006). For example, the Masdar Institute of Science and Technology in Abu Dhabi, which offers research-based programmes at the graduate level, has a collaborative agreement with MIT (Reiche, 2010b). A recent example of R&D is the initiative of the King Abdullah University of Science and Technology (KAUST) in Saudi Arabia. Here, the University aims to reinforce further its R&D through embarking on a five-year research collaboration plan with the Stuttgart-based Centre for Solar Energy and Hydrogen Research Baden Württemberg (ZSW), a German solar energy institute, to examine the performance of PVs under desert conditions (Hall, 2014). Another example is the R&D collaboration agreement between The King Abdullah City for Atomic and Renewable Energy (K.A.CARE) and The Fraunhofer Institute for Solar Energy Systems (Fraunhofer-ISE) (Osborne, 2014). Ultimately, these projects will also raise public awareness around renewable solar energy.

At the moment, the main drawback of commercially available, silicon-based PVs is the initial cost, which impedes their wide domestic adoption in the absence of subsidies. However, proactive research work is being conducted on PVs based on organic semiconductor polymers, which are a form of plastic. Plastic PVs have the advantage of low cost, ease of fabrication, mechanical flexibility and **the possibility of mass production using a roll-to-roll printing technique** over their silicon-based counterpart, which would ease their portability and make them more favourable for rural areas. Therefore, plastic PVs have attracted a vast amount of interest from researchers around the globe due to their application in optoelectronic devices, which has resulted in a rapid progression in their efficiency (Deschler *et al.*, 2014, Hao *et al.*, 2014, Yeh and Yeh, 2013). However, the efficiency of organic polymer solar cells is still beyond the commercial standard and much remains to be researched regarding the basic mechanisms that ultimately control the efficiency of the light-to-electricity conversion. In response, researchers around the globe are conducting large-scale studies on plastic solar cells to meet their commercialisation needs. Here arises the importance of grants and incentives available for R&D in aiding the efforts to achieve the goal of fabricating efficient, low-cost as well as environmentally friendly PVs. Furthermore, there are some researchers exploring the possibility of recycling of commercially available PVs which will help reducing the cost of PVs, saving the available resources as well as decreasing the environmental impacts (Fthenakis *et al.*, 2009).

The incorporation of solar energy appliances constitutes one of the bases of green buildings. Contrary to old, traditional buildings, current built environments are not energy efficient. This focuses attention on houses and the built environment and the call for producing sustainable architecture (Taleb and Sharples, 2011).One of the policy measures in the UAE aimed at sustainable architecture is the green building code, which has been introduced in Dubai for all new buildings from 2009. This step evolved following the launch of Masdar City in Abdu Dhabi, indicating the progressive revision and transfer of policies between neighbouring emirates in favour of those supporting the environment (Reiche, 2010a). Furthermore, the current built environment can be improved through the integration of solar appliances such as PVs and solar water heating into buildings. For example, a study in Bahrain showed that a building-integrated photovoltaic (BIPV) would produce up to 30% of the total electricity expended in the built environment (Bachellerie, 2012). Also, a study on a residential building in Saudi Arabia estimated that the incorporation of eight solar PV panels on the roof of the building would generate 10% of the household electricity demand (Taleb and Sharples, 2011). A recent study in the UAE on a selected building shows the potential to decrease energy consumption by utilising a solar water heater (Taleb and Al-Saleh, 2014). Furthermore, energy efficiency standards should be set up for all the imported electrical appliances and vehicles, which will help in reducing the domestic intake of energy and also raise public awareness about energy conservation.

Current GCC programmes to incorporate solar as a means of diversifying the energy portfolio have exacerbated a demand for well-trained, highly skilled human capital. This demand is expected to increase in future as plans expand and more projects take place. For example, the rapid growth of the renewable energy field is expected to cause expansion in the employment market in the US and Germany (Ulrike Lehr, 2012, Wei *et al.*, 2010). Similarly, it is estimated that the renewable energy field will create about 180,000 jobs in the Middle East by 2050 on the basis that manufacturing is taking place locally (van der Zwaan *et al.*, 2013). The new job vacancies include technicians, teachers, engineers, and scientists who are needed in the design, installation, and maintenance of solar appliances. Additionally, there will also be job openings related to energy policy, policy analysis, law, sales and marketing that will assist in trading, and evaluation and formulation of the relevant policies and legislations. Governments will need to quantitatively plan ahead in anticipation of the high demand on human resources in relation to projected jobs in the solar energy sector and associated educational and training programmes and public policy (Alnaser and Flanagan, 2007, *Building a Sustainable Energy Future: U.S. Actions for an Effective Energy Economy Transformation*, 2009, Haag *et al.*, 2012, Wei *et al.*, 2010). Therefore, it would be advantageous for the GCC countries to dedicate efforts and funds to invest in youth capital as human resources to help meet future demand. Currently, despite the efforts of the GCC countries to meet the surge in demand of energy through the implementation of a non-fossil fuel energy mix strategy, a comprehensive educational and training programme which addresses the various technical and nontechnical aspects related to solar energy and promotes its evolution is still absent. Training courses related to solar energy are included within disciplines such as science, material science or engineering which are unable to provide the growing demand for various levels of solar energy professionals. Therefore, the capability to supply the required efficient training should be established in the GCC countries to fulfil the need for skilled solar energy experts. To overcome this shortage and in response to such specific demand, two experiences are presented below as examples of relevant emerging programmes from leading universities.

The first example is a graduate programme in the United States introduced by the University of Arizona in 2011 following a survey of a large number of solar energy companies to define the required and desired skills of prospected employees (Phelan and Dada, 2012).The programme provides the students with scientific knowledge, through technical and nontechnical coursework. In addition, the programme offers a culminating project that is uniquely supervised by two mentors – an academic mentor and a professional one from the solar energy field. Such dual-supervision ensures that the project takes into account real-world implications and suggestions, whilst it also helps the student to establish solid network linkages in both sectors. (The professional can be from the government, industry or the non-profit sector.) Furthermore, a key part of the programme is energy policy knowledge, for which the student spends one week of face-to-face interaction with policy- and decision-makers. Such real work experience aims to prepare students to be able to communicate and implement energy policy at both federal and international levels. The programme has proved to be successful and is attracting a growing number of both national and international students since its commencement.

The second example is from Australia, which has a long record of establishing solar energy education as a new discipline. The first step towards solar energy education was in 1988, when a Certificate IV in Renewable Energy was introduced by the Brisbane and North Point Institute of Technical and Further Education (TAFE) and the programme was supported by the Business Council for Sustainable Energy (BCSE) (Jennings, 2009, Jennings and Lund, 2001). An interdisciplinary programme was launched at Murdoch University, Australia, as a Diploma in 1992 and was followed by a two-year Master's programme in 1998 (Jennings and Lund, 2001). The programme was specially planned to meet the principles of ecological sustainable development (ESD) in order to provide the participants with awareness of the philosophy, training and knowledge surrounding ESD. The modules in the programme comprehensively span a wide range of topics related to renewable solar energy as well as taking into account the environmental, social and economic aspects associated with the solar energy field (*The global partnership for environment and development: a guide to Agenda 21*, 1992).

The programme's strong and extensive module content together with its flexibility in terms of availability as part-time or via distance learning has led to its success and it has attracted both national and international students. Graduates have successfully engaged in solar energy occupations in both research and industry or in the public sector related to energy or policy. The University also provides a course in Environmental Architecture, which delivers the fundamentals of energy-efficient building and lines up well with the principles of ESD. The success of the programme encouraged Massey University in New Zealand to adopt some of the modules. Furthermore, in an attempt to react to the mounting requests for renewable energy professionals, in 2002 Murdoch University launched an undergraduate programme in Energy Studies, renamed later as Sustainable Energy Studies. Also, in 2001 the engineering department introduced a renewable energy course following a call from the Western Power Cooperation to adopt renewable energy in power generation. Moreover, the University is providing several short courses to keep the professionals in the solar energy field up-to-date with developments and knowledge, as well as providing information online to reinforce public knowledge, familiarity and awareness (Jennings, 2009).

The two examples discussed above demonstrate that a specially designed solar energy educational and training programme will help in delivering professionals who are accurately equipped with the skills and knowledge needed to achieve the requirements of the job market both in the current and future stages in the entire Gulf region in accordance with the call addressed by decision-makers, consultants and contractors in a survey performed in Bahrain (Alnaser and Flanagan, 2007). Although the request for such special programmes was presented by academics in the Gulf region, who put forward a designed training programme for the region, no such programmes have yet commenced (Alnaser and Al-Karaghouli, 2001). However, the uptake of launching these programmes is anticipated to be emulated in the GCC countries to fulfil the call for jobs relevant to solar energy created by new solar energy projects and the expansion of the already existing ones (Reiche, 2010a).

Moreover, the availability of these programmes will assist in nurturing the public awareness about solar energy and the economic and environmental benefits that can be gained from its widespread adoption. It is also suggested by a study performed on the solar energy policy of Malaysia to introduce solar energy into the science curriculum and activities at primary and secondary school levels to help in raising public awareness – a step which is already taking place in the GCC states (Mekhilef *et al.*, 2012, Patlitzianas *et al.*, 2006).

7. Conclusion

That GCC countries have begun to accelerate exploitation of the abundant solar energy freely available for electricity generation is evidenced by the establishment of new projects and the expansion of already functioning ventures. For these countries, the well-timed adoption of these initiatives to employ solar energy for electricity production as an alternative to fossil fuels is auspicious for two reasons. First, it will help them to meet the growing demand for heavily subsidised fossil fuels energy: energy subsidies are actually exerting massive strains on national budgets as well as raising concerns about export capacity; however, they are reckoned to be a politically and economically sensitive topic.

Second, employing solar energy, which is a clean and sustainable resource, will allow the GCC countries to comply with international obligations associated with the ratification of Kyoto and the UNFCCC. Additionally, the GCC countries' use of solar energy coincides well with the sustainability pillars. Therefore, they have implemented an energy portfolio into their future economic vision in which solar energy is proactively stated to be utilised. Their initiatives towards the employment of solar energy are evidently endorsed and gradually diffused between the GCC countries. Looking at the energy policy in the Gulf, one finds that several measures are being conducted, primarily in relation to R&D and utility-scale deployment. However, a coherent, integrated comprehensive plan of the

GCC countries' governmental endeavours to employ solar energy in power production is still absent. This chapter has drawn attention to two drawbacks in the energy policy of the GCC countries: first, their lack of programmes facilitating the domestic diffusion of solar appliances and, second, the short supply of purposely designed renewable (including solar) energy educational and training programmes that would prepare the required competent human resources.

There is a crucial need to formulate and implement a strategic energy policy that encourages the domestic distribution of solar appliances and sustainable architecture guidelines. Domestic dissemination should be encouraged by introducing subsidy schemes to make PVs cost-effective and competitive with conventional fossil fuel electricity from the consumer's perspective. The chapter has suggested that the introduction of subsidies will help the domestic placement of solar appliances and so make the available buildings more environmentally friendly. It will also help the GCC countries to reduce the current burden on fossil fuels due to domestic electricity consumption as well as to increase export capacity and revenues. Conserving their endowment of fossil fuel together with the prospect of exporting the surplus solar energy will ultimately enable and support the GCC countries to maintain their leading position as global economic powers.

Today, the available programmes are limited or revolve around R&D, graduate level training, or courses that are implemented within other disciplines. Therefore, in addition to those programmes, the chapter encourages the GCC countries to develop specifically designed solar energy-oriented training and educational programmes that are currently lacking. These comprehensive education and training programmes will enable them to develop skilled human capital that conforms to the current and projected demands for professionals. To pursue this recommendation, this chapter has discussed two programmes for fulfilling the short supply of such training. These programmes will also have indirect impacts in public awareness raising. Additionally, this chapter also advises the insertion of energy efficiency standards for all the imported electrical appliances and vehicles, and the establishment of measures that promote sustainable architecture, such as the green code building which at present is in operation only in Dubai.

References

Ahmad, A. & Ramana, M. V. 2014. *Too costly to matter: Economics of nuclear power for Saudi Arabia.* Energy, 69, 682-694.

Al-Jayousi, M. 2012. *Bahrain announces pioneering solar energy project.* Al-Shorfa.com, 30th May, 2012. Available from: http://al-shorfa.com/en_GB/articles/meii/features/main/2012/05/30/feature-03 [Accessed 25th September, 2014].

Al-Jayousi, M. 2013. *Solar plant planned for Bahrain's Hawar Islands.* Al-Shorfa.com, 3rd May, 2013. Available from: http://al-shorfa.com/en_GB/articles/meii/features/2013/05/03/feature-03 [Accessed 3rd June, 2014].

Alnaser, N. W. & Flanagan, R. 2007. *The need of sustainable buildings construction in the Kingdom of Bahrain.* Building and Environment, 42, 495-506.

Alnaser, W. E. 1995. *Renewable energy resources in the state of Bahrain.* Applied Energy, 50, 23-30.

Alnaser, W. E. & Al-Karaghouli, A. A. 2001. *Training package in new energy sources and technology for the Arab world.* Renewable Energy, 24, 373-387.

Alnaser, W. E. & Aldudiafa, H. S. 1990. *Calculation of the global, diffused, and direct solar radiation in Bahrain.* Solar & Wind Technology, 7, 309-311.

Al-Qahtani, H. 1996. *Feasibility of utilizing solar energy to power reverse osmosis domestic unit to desalinate water in the state of Bahrain.* Renewable Energy, 8, 500-504.

Bachellerie, I. J. 2012. Renewable energy in the GCC countries: resources, potential, and prospects. Available from: http://library.fes.de/pdf-files/bueros/amman/09008.pdf [Accessed 4th May, 2014].

Bahrain boils in summer power cut. 2004. The BBC, 23rd August, 2004. Available from: http://news.bbc.co.uk/1/hi/world/middle_east/3590798.stm [Accessed 28th September, 2014].

Bahrain World Trade Center. 2014. Available from: http://en.wikipedia.org/wiki/Bahrain_World_Trade_Center [Accessed 23rd August, 2014].

Bhutto, A. W., Bazmi, A. A., Zahedi, G. & Klemeš, J. J. 2014. *A review of progress in renewable energy implementation in the Gulf Cooperation Council countries.* Journal of Cleaner Production, 71, 168-180.

Bollier, S. 2014. *The Gulf's bright solar-powered future.* Aljazeera.com, 23rd January, 2014. Available from: http://www.aljazeera.com/indepth/features/2014/01/gulf-bright-solar-powered-future-201412363550740672.html [Accessed 8th June, 2014].

Chang, P.-L., Ho, S.-P. & Hsu, C.-W. 2013. *Dynamic simulation of government subsidy policy effects on solar water heaters installation in Taiwan.* Renewable and Sustainable Energy Reviews, 20, 385-396.

Clercq, G. D. 2013. *Saudi Arabia hopes to export solar electricity to Europe* Reuters.com, 11th April, 2013. Available from: http://uk.reuters.com/article/2013/04/11/saudi-solar-europe-idUKL5N0CY2SX20130411 [Accessed 7th June, 2014].

Clover, I. 2013. *Qatar to install utility-scale reservoir rooftop solar panels.* Pv-magazine.com, 20th November, 2013. Available from: http://www.pv-magazine.com/news/details/beitrag/qatar-to-install-utility-scale-reservoir-rooftop-solar-panels_100013499/#ixzz34Gr77Dtp [Accessed 8th June, 2014].

Clover, I. 2014a. *Qatar Solar Energy opens 300 MW PV factory in Qatar.* Pv-magazine.com, 10th June, 2014. Available from: http://www.pv-magazine.com/news/details/beitrag/qatar-solar-energy-opens-300-mw-pv-factory-in-qatar_100015352/#ixzz34KWdzecM [Accessed 10th June, 2014].

Clover, I. 2014b. *Yingli mulls bid for 100 MW solar plant in Dubai.* Pv-magazine.com, 28th May, 2014. Available from: http://www.pv-magazine.com/news/details/beitrag/yingli-mulls-bid-for-100-mw-solar-plant-in-dubai_100015252/#ixzz34LBKVLTw [Accessed 30th May, 2014].

Dargin, J. 2010. *The GCC in 2020: Resources for the future.* Economist Intelligence Unit. Available from: http://graphics.eiu.com/upload/eb/GCC_in_2020_Resources_WEB.pdf [Accessed 4th May, 2014].

Dawoud, M. A. & Mulla, M. M. A. 2012. *Environmental Impacts of Seawater Desalination: Arabian Gulf Case Study.* International Journal of Environment and Sustainability, 1, 22-37.

Deschler, F., Price, M., Pathak, S., Klintberg, L. E., Jarausch, D.-D., Higler, R., Hüttner, S., Leijtens, T., Stranks, S. D., Snaith, H. J., Atatüre, M., Phillips, R. T. & Friend, R. H. 2014. *High Photoluminescence Efficiency and Optically Pumped Lasing in Solution-Processed Mixed Halide Perovskite Semiconductors.* The Journal of Physical Chemistry Letters, 5, 1421-1426.

Emerging Nuclear Energy Countries. 2014. Available from: http://www.world-nuclear.org/info/Country-Profiles/Others/Emerging-Nuclear-Energy-Countries/ [Accessed 25th September, 2014].

Fattouh, B. & El-Katiri, L. 2013. *Energy subsidies in the Middle East and North Africa.* Energy Strategy Reviews, 2, 108-115.

Fthenakis, V. 2009. *Sustainability of photovoltaics: The case for thin-film solar cells.* Renewable and Sustainable Energy Reviews, 13, 2746-2750.

Fthenakis, V., Mason, J. E. & Zweibel, K. 2009. *The technical, geographical, and economic feasibility for solar energy to supply the energy needs of the US.* Energy Policy, 37, 387-399.

H.E. The Ministry of Energy signs a Contract for Conducting Feasibility Study for Renewable Energies and Supervision of Establishment of a Pilot Power Plant [Online]. 2011. Bahrain. Available from: http://www.mew.gov.bh/default.asp?action=article&ID=1107 [Accessed 4th May, 2014].

Haag, S., Pasqualetti, M. & Manning, M. 2012. *Industry perceptions of solar energy policy in the American southwest.* Journal of Integrative Environmental Sciences, 9, 37-50.

Hall, M. 2014. *German center to work with Saudi university.* Pv-magazine.com, 5th May, 2014. Available from: http://www.pv-magazine.com/news/details/beitrag/german-center-to-work-with-saudi-university_100014979/#ixzz34FuQoM21 [Accessed 9th June, 2014].

Hao, F., Stoumpos, C. C., Cao, D. H., Chang, R. P. H. & Kanatzidis, M. G. 2014. *Lead-free solid-state organic-inorganic halide perovskite solar cells.* Nature Photonic, advance online publication.

Hepbasli, A. & Alsuhaibani, Z. 2011. *A key review on present status and future directions of solar energy studies and applications in Saudi Arabia.* Renewable and Sustainable Energy Reviews, 15, 5021-5050.

International Energy Agency (IEA). 2013. *CO_2 Emissions from fuel combustion - highlights.* International Energy Agency (IEA) Publication. Available from: http://www.iea.org/publications/freepublications/publication/CO2EmissionsFromFuelCombustionHighlights2013.pdf [Accessed 28th September, 2014].

Jablonski, S., Tarhini, M., Touati, M., Gonzalez Garcia, D. & Alario, J. 2012. *The Mediterranean Solar Plan: Project proposals for renewable energy in the Mediterranean Partner Countries region.* Energy Policy, 44, 291-300.

Jennings, P. 2009. *New directions in renewable energy education.* Renewable Energy, 34, 435-439.

Jennings, P. & Lund, C. 2001. *Renewable energy education for sustainable development.* Renewable Energy, 22, 113-118.

Kemp, J. 2014. *COLUMN-Money to burn: OPEC's wasteful energy subsidies.* Reuters.com, 16th May, 2014. Available from: http://www.reuters.com/article/2014/05/16/opec-fuel-subsidies-idUSL6N0O231C20140516 [Accessed 8th June, 2014].

Krane, J. 2012. *Energy Policy in the Gulf Arab States: Shortage and Reform in the World's Storehouse of Energy. In:* 31st USAEE/IAEE North American Conference:"Transition to sustainable Energy era: Opportunities and Challenges", Austin, Texas. United States Association for Energy Economics.

Kuwait's first PV plant underpinned by skytron energy's monitoring solution. 2013. Pv-magazine.com, 25th November, 2013. Available from: http://www.pv-magazine.com/services/press-releases/details/beitrag/kuwaits-first-pv-plant-underpinned-by-skytron-energys-monitoring-solution_100013539/#ixzz34FzsuJgK [Accessed 10th June, 2014].

Kuwait may face electricity blackouts 2013. Arabianbusiness.com, 12th March, 2013. Available from: http://www.arabianbusiness.com/kuwait-may-face-electricity-blackouts-492799.html [Accessed 7th June, 2014].

Lahn, G. & Stevens, P. 2011. *Burning Oil to Keep Cool: The Hidden Energy Crisis in Saudi Arabia,* London, Chatham House (The Royal Institute of International Affairs). Available from: http://www.chathamhouse.org/publications/papers/view/180825 [Accessed 1st December, 2011].

Lahn, G., Stevens, P. & Preston, F. 2013. *Saving Oil and Gas in the Gulf,* London, Chatham House (The Royal Institute of International Affairs).

Largest solar powered desalination plant to be built in Al-Khafji. 2013. 12th December, 2013. Available from: http://www.arabnews.com/news/491201 [Accessed 22nd June, 2014].

Law, B. 2012. *Gulf states face hard economic truth about subsidies.* The BBC, 18th December, 2012. Available from: http://www.bbc.co.uk/news/world-middle-east-20644964 [Accessed 8th June, 2014].

Major boost for renewable energy in Bahrain. 2014. Bahrain News Agency, 11th November, 2014. Available from: http://www.bna.bh/portal/en/news/639675 [Accessed 5th December, 2014].

Mayton, J. 2012. *Bahrain Announces 5 MW Solar Power Entrance.* Greenprophet.com, 14th October, 2012. Available from: http://www.greenprophet.com/2012/10/bahrain-5mw-pv-power-plant/#sthash.RuO62VSW.dpuf [Accessed 8th June, 2014].

Mclauchlan, L. & Mehrubeoglu, M. 2010. *A survey of green energy technology and policy. In:* 2010 IEEE Green Technologies.

Mekhilef, S., Safari, A., Mustaffa, W. E. S., Saidur, R., Omar, R. & Younis, M. a. A. 2012. *Solar energy in Malaysia: Current state and prospects.* Renewable and Sustainable Energy Reviews, 16, 386-396.

Meza, E. 2013. *Kuwait opens bidding for first phase of 2 GW clean energy park*. Pv-magazine.com, 13th June, 2013. Available from: http://www.pv-magazine.com/news/details/beitrag/kuwait-opens-bidding-for-first-phase-of-2-gw-clean-energy-park_100011708/#ixzz34KscvYw4 [Accessed 7th June, 2014].

Meza, E. 2014. *SunEdison eyes PV manufacturing complex in Saudi Arabia*. Pv-magazine.com, 2nd February, 2014. Available from: http://www.pv-magazine.com/news/details/beitrag/sunedison-eyes-pv-manufacturing-complex-in-saudi-arabia_100014119/#axzz34Fpfx8aj [Accessed 9th June, 2014].

National Science Board. 2009. *Building a Sustainable Energy Future: U.S. Actions for an Effective Energy Economy Transformation*. National Science Foundation. Available from: <http://www.nsf.gov/pubs/2009/nsb0955/index.jsp> [Accessed 22nd September, 2014].

Nouh, M. 2008. An overview for the water resources of the United Arab Emirates. *The 1st Technical Meeting of Muslim Water Researchers Cooperation (MUWAREC)* [Online]. Available from: http://www.ukm.my/muwarec/ProceedingMuwarec08/3-MamdouhNouh_UAE%2011nov08.pdf [Accessed 26th September, 2014].

Nuclear power in Saudi Arabia. 2014. World Nuclear Association (WNA). Available from: http://www.world-nuclear.org/info/Country-Profiles/Countries-O-S/Saudi-Arabia/ [Accessed 26th September, 2014].

Omari, A. A. 2012. *Nuclear power plan postponed*. Gulf Daily News, 17th October, 2012. Available from: http://www.gulf-daily-news.com/source/XXXV/211/pdf/page13.pdf [Accessed 7th June, 2014].

OPEC member Qatar says to start embracing solar energy. 2012. Reuters.com, 1st December, 2012. Available from: http://www.reuters.com/article/2012/12/01/us-climate-qatar-talks-idUSBRE8B00A720121201 [Accessed 2nd June, 2014].

Osborne, M. 2014. *Fraunhofer ISE to team with Saudi Arabian KA.CARE on renewables R&D*. www.pv-tech.org, 22nd January, 2014. Available from: http://www.pv-tech.org/news/fraunhofer_ise_to_team_with_saudi_arabian_k.a.care_on_renewables_rd [Accessed 8th June, 2014].

Patlitzianas, K. D., Doukas, H. & Psarras, J. 2006. *Enhancing renewable energy in the Arab States of the Gulf: Constraints & efforts*. Energy Policy, 34, 3719-3726.

Phelan, P. & Dada, K. 2012. *Solar energy education: The professional science master's degree in solar energy engineering & commercialization at arizona state university*. *In:* World Renewable Energy Forum, WREF 2012, Including World Renewable Energy Congress XII and Colorado Renewable Energy Society (CRES) Annual Conferen. 3455-3459.

Rafique, M. 2012. *Thai EHC delegation visits Hawar Island*. Twenty Four Seven News.com 20th October, 2012. Available from: http://www.twentyfoursevennews.com/bahrain-news/thai-electricity-holding-company-visits-hawar-island/ [Accessed 4th June, 2014].

Reiche, D. 2010a. *Energy Policies of Gulf Cooperation Council (GCC) countries—possibilities and limitations of ecological modernization in rentier states*. Energy Policy, 38, 2395-2403.

Reiche, D. 2010b. *Renewable Energy Policies in the Gulf countries: A case study of the carbon-neutral "Masdar City" in Abu Dhabi*. Energy Policy, 38, 378-382.

Riyami, A. a. A. 2013. *Pact signed for Oman's first solar energy plant*. Oman Observer, 14th November, 2013. Available from: http://main.omanobserver.om/?p=30487 [Accessed 6th December, 2014].

Saadi, D. 2014. *UAE renewable energy projects bring the future into view*. The National, 18th January, 2014. Available from: http://www.thenational.ae/business/industry-insights/energy/uae-renewable-energy-projects-bring-the-future-into-view#ixzz34L8QPZ2N [Accessed 8th June, 2014].

Sarzynski, A., Larrieu, J. & Shrimali, G. 2012. *The impact of state financial incentives on market deployment of solar technology*. Energy Policy, 46, 550-557.

Shams solar power station [Online]. Available from: http://en.wikipedia.org/wiki/Shams_solar_power_station [Accessed 26th September, 2014].

Sharples, S. & Radhi, H. 2013. *Assessing the technical and economic performance of building integrated photovoltaics and their value to the GCC society*. Renewable Energy, 55, 150-159.

Solangi, K. H., Islam, M. R., Saidur, R., Rahim, N. A. & Fayaz, H. 2011. *A review on global solar energy policy*. Renewable and Sustainable Energy Reviews, 15, 2149-2163.

Solar-powered street lighting explored 2014. Bahrain News Agency, 18th March, 2014. Available from: http://www.bna.bh/portal/en/news/609117 [Accessed 4th June, 2014].

Stuart, B. 2011. *Phoenix Solar to build PV park for Saudi oil company*. Pv-magazine.com, 14th February, 2014. Available from: http://www.pv-magazine.com/news/details/beitrag/phoenix-solar-to-build-pv-park-for-saudi-oil-company_100002222/#ixzz3EQeD7oZl [Accessed 26th September, 2014].

Swift, K. D. 2013. *A comparison of the cost and financial returns for solar photovoltaic systems installed by businesses in different locations across the United States*. Renewable Energy, 57, 137-143.

Taleb, H. & Al-Saleh, Y. 2014. *Applying energy-efficient water heating practices to the residential buildings of the United Arab Emirates*. International Journal of Environmental Sustainability, 9, 35-51.

Taleb, H. M. & Sharples, S. 2011. *Developing sustainable residential buildings in Saudi Arabia: A case study*. Applied Energy, 88, 383-391.

UAE nuclear plant unit 3 work starts. 2014. Trade Arabia Business News Information, 24th September, 2014. Available from: http://www.tradearabia.com/news/OGN_266396.html [Accessed 25th September, 2014].

UAE ranks third in the world for total solar power capacity and investment. 2014. The National, 4th June, 2014. Available from: http://www.thenational.ae/business/shams-1-solar-power-plant#ixzz34L0E95hi [Accessed 11th June, 2014].

Ulrike Lehr, C., Dietmaredler. 2012. *Green jobs? Economic impacts of renewable energy in Germany*. Energy Policy, 47, 358–364.

United Nations Conference on Environment & Development (UNCED). 1992. *The global partnership for environment and development: a guide to Agenda 21*. Available from: http://sustainabledevelopment.un.org/content/documents/Agenda21.pdf [Accessed 28th September, 2014].

United Nations Environment Programme (UNEP). 2010. *Green economy report: a preview*. Available from: http://www.unep.org/pdf/GreenEconomyReport-Preview_v2.0.pdf [Accessed 28th September, 2014].

Van Der Zwaan, B., Cameron, L. & Kober, T. 2013. *Potential for renewable energy jobs in the Middle East*. Energy Policy, 60, 296-304.

Wei, M., Patadia, S. & M.Kammena, D. 2010. *Putting renewables and energy efficiency to work: How many jobs can the clean energy industry generate in the US?* Energy Policy, 38, 919–931.

Whittell, G. 2014. *The future's bright if we can trap the Saharan sun*. The Times, 7th June, 2014. Available from: http://www.thetimes.co.uk/tto/opinion/columnists/article4111699.ece [Accessed 8/6/2014].

Willis, B. 2012. *Dubai looks to renewables to meet future energy demand*. PV Tech, 6th November, 2012. Available from: http://www.pv-tech.org/news/dubai_looks_to_renewables_to_meet_future_energy_demand [Accessed 8th June, 2014].

Woods, L. 2013. *Shell subsidiary celebrates six months of providing Aramco with solar energy*. PV Tech, 28th August, 2013. Available from: http://www.pv-tech.org/news/shell_subsidiary_celebrates_six_months_of_providing_aramco_with_solar_e [Accessed 8th June, 2014].

Woods, L. 2014. *Bahrain 5MW solar plant complete*. PV Tech, 25th June, 2014. Available from: http://www.pv-tech.org/news/bharian_5mw_solar_plant_complete [Accessed 23rd September,2014].

Wynn, G. 2014. *Pressure growing for OPEC countries to cut energy subsidies* [Online]. Available from: http://www.rtcc.org/2014/05/09/pressure-growing-for-opec-countries-to-cut-energy-subsidies/# [Accessed 22nd September, 2014].

Yee, A. 2013. *Saudi Arabia to seek bids for its first nuclear reactor*. The National, 12th Novermber, 2013. Available from: http://www.thenational.ae/business/industry-insights/energy/saudi-arabia-to-seek-bids-for-its-first-nuclear-reactor [Accessed 7th June, 2014].

Yeh, N. & Yeh, P. 2013. *Organic solar cells: Their developments and potentials*. Renewable and Sustainable Energy Reviews, 21, 421-431.

Zhai, P. 2013. *Analyzing solar energy policies using a three-tier model: A case study of photovoltaics adoption in Arizona, United States*. Renewable Energy, 57, 317-322.

About the Contributors

Mayami Abdulla is currently a researcher at Cambridge University. She holds a PhD in Physics of plastic solar cells from the University of Cambridge. Abdulla has published papers in the area of organic optoelectronics. She worked as an academic at the University of Bahrain and a science curricula specialist at the Ministry of Education in Bahrain. Abdulla has participated in several specialised conferences. Her interest is solar cells, energy policy, sustainable development and science education.

John Anthony Allan is based at King's College London. Allan [BA Durham 1958, PhD London 1971] heads the London Water Research Group at King's College London and SOAS. He specialises in the analysis of water resources in semi-arid regions and on the role of global systems in ameliorating local and regional water deficits. Allan provides advice to governments and agencies especially in the Middle East on water policy and water policy reform. His ideas on water security are set out in The Middle East water question: hydropolitics and the global economy [2001] and in a recent book entitled Virtual water [2011]. He has also addressed recent water and land grabbing in the co-edited Handbook on land and water grabbing in Africa [2012].

Yousuf Hamad Al-Balushi is an economist with a track record of over 18 years of professional experience at the Central Bank of Oman and the Supreme Council for National Development and Planning of Oman. Al-Balushi, appointed as Foreign Direct Investment (FDI) Statistics Advisor under the technical assistance program of the International Monetary Fund (IMF). He taught at King's College London, modules relating to Foreign Trade and Political Economy of the Middle East: Theory and Practice. Currently Al-Balushi is undertaking PhD at KCL, University of London focusing on the impact of FDI on the efficiency of the private sector and the economic development process in Oman. He is regularly give lectures and speeches at the University of Cambridge, the Central Bank of Oman and Sultan Qaboos University on topics including monetary policy, foreign trade, foreign direct investment and economic development.

Mhamed Biygautane is a research associate in the Government and Public Management Program at the Mohammed Bin Rashid School of Government (MBRSG), specializing in talent management, training in the public sector, governance, public management/ administration, and sustainable economic growth for the UAE and wider Gulf Cooperation Council's (GCC) region. He also serves as a Middle East and North Africa (MENA) Expert in the European Geopolitical Forum where he provides strategic advice on the economic development of the GCC and MENA region. Biygautane has published more than 70 peer-reviewed studies on political economy of development, knowledge management and training in the public sector, modernization and the reform of public sector organizations. He also contributed to case studies' series that the MBRSG has co-published with the World Bank, and actively contributes to Oxford Analytica, the Fair Observer and Al Arabiya English.

David Bryde is professor of Project Management in the Built Environment & Sustainable Technologies (BEST) Research Institute, Liverpool John Moores University (LJMU). He brings a management science/ social science perspective to the topic of sustainability. Working with Al-Rasheed and Mouzughi he is investigating attitudes and behaviours towards sustainability among key stakeholders in Saudi Arabia. His work has also encompassed the topics of sustainable project management and sustainable procurement. Bryde is widely published, with over 80 journal papers, research monographs, book chapters, conference presentations, invited guest lecturers/presentations, expert interviews and articles.

Justin Dargin is an Energy Scholar at the University of Oxford. He was a former Research Fellow with The Dubai Initiative at Harvard University, where he won a Harvard award for his research into the MENA energy/power sector. Dargin was also an Aramco-OIES Senior Fellow and worked in the legal department at the Organization of Petroleum Exporting Countries, where he advised the senior staff as to the implications of several multilateral initiatives with the WTO and UN. Dargin has also advised some of the world's largest international and national oil companies as to strategic investment policy in the MENA region.

Samantha Dobbie is a postgraduate research student at the University of Southampton. Belonging to both the Institute for Complex Systems Simulation (ICSS) and the Centre for Environmental Sciences (CES), her interests lie in complex social-ecological systems. At the moment Dobbie is exploring the use of simulation tools to better understand the multi-dimensional nature of food security. In particular, she is investigating the potential of agent-based modeling to unravel the complex social, ecological and political factors affecting food availability, access, utilisation and stability.

James G. Dyke holds a PhD in Informatics from the Centre for Computational Neuroscience & Robotics of the University of Sussex. Up to 2011 Dyke was postdoctoral research scientist in the Biospheric Theory & Modelling Group of the Max Planck Institute for Biogeochemistry in Jena. Now he is a lecturer in Complex Systems (Simulation) at the University of Southampton. Besides Dyke is the program director and co-chair for a new interdisciplinary Masters in Sustainability.

Nilly Kamal Elamir is Egyptian researcher at the fields of Environmental Politics and their relation to development and Asian Affairs. She has earned her PhD from School of Economics and Political Science of Cairo University on the topic of Environmental Security Politics: Comparative Study between US and Japan in 2010, and MA degree from the same school on "The Concept of Trans-regionalism: Case of Asia Europe Meeting (ASEM)". Elamir was visiting researcher in 2007 and 2008 at School of Asia – Pacific Studies of Waseda University in Japan. In addition to paper submission in international conferences and publishing research papers, she writes articles on Arab newspaper sometimes such as Almasy Elyoum (Egypt) and Alkhaleej (UAE). Elamir focuses also on community based development and environmental based business (SMEs) as an instrument for combating poverty and improving live quality in underprivileged communities.

Paul Joyce is visiting professor of Public Services, Leadership & Strategy at Birmingham City Business School. His expertise in research and consultancy is in public management, the management of reform and modernization of the public sector, especially developments in strategic management, leadership and change management. Joyce's work in this area includes: *Strategic Management in the Public Sector* (2015); *Strategic Leadership in the Public Services* (2011); *Lessons in Leadership: Meeting the Challenges of Public Services Management* (2005) (co-authored with Eileen Milner); *Strategy in the Public Sector: A guide to Effective Change Management* (2000); and *Strategic Management for the Public Services* (1999). He is the co-chair of a permanent study group of the European Group for Public Administration (EGPA) that looks at strategic management in government.

Martin Keulertz is a research fellow in the Department for Agricultural Economics at Humboldt University of Berlin. From 2014 to early 2015, he has worked as a post-doctoral research associate in the Department for Agricultural and Biological Engineering at Purdue University in Indiana (United States). Keulertz is part of the core Water-Energy-Food Nexus team around Rabi Mohtar (Texas A&M). In 2012, Water

International selected his co-authored article on the role of corporates in global 'virtual water' trade as paper of the year. He further led the editing process of the Handbook of Land and Water Grabs in Africa: foreign direct investment and food and water security published by Routledge in 2013. Keulertz is a member of the London Water Research Group chaired by Tony Allan of King's College London. He worked for Dresdner Bank in Germany, consulted energy utility companies, development agencies and the United Nations on issues around the green economy.

Latifa Al-Khalifa is a research assistant at the Bahrain Center for Strategic, International and Energy Studies, DERASAT, specializing in Strategic Studies. Her primary research interests include foreign policy decision-making, foreign policy change, and domestic politics and foreign policy interaction. Prior to joining DERASAT, Al-Khalifa earned her M.A. in International Relations from Webster Graduate School in London, a B.A. with a Double-Major in Politics and International Studies from Monash University and a Diploma in Communications and Human Behavior from Monash College in Melbourne, Australia.

Jerry Kolo, PhD, is a professor in the Master of Urban Planning Program at the American University of Sharjah (AUS) in the United Arab Emirates. He joined AUS in 2006 from Florida Atlantic University in Fort Lauderdale, Florida, USA, where he was a professor of urban planning and the founder and director of an applied research and community outreach *Center for Urban Redevelopment and Empowerment* (CURE). Kolo's teaching and research areas are sustainable community planning processes, land use planning and tourism development. He has extensive consultancy experience in public-sector planning and community capacity building.

Rachael McDonnell is a scientist at the International Center for Biosaline Agriculture (ICBA) in Dubai and Senior Visiting Research Associate at the University of Oxford's School of Geography and Environment. At ICBA she is Head of the Climate Change Modelling and Adaptation Section leading new projects and initiatives to understand the impacts on water and food security of future climate conditions, particularly extreme events such as droughts. McDonnell continues to work in the areas of water policy and governance exploring complex issues associated with managing scare resources in the politically fractured and evolving environments of the Middle East and North Africa.

Yusra Mouzughi is a principal lecturer at Liverpool Business School and the programme leader for doctoral programmes. Mouzughi's research is cross disciplinary focusing on a broad range of sustainability issues including general attitudes towards sustainability as well as the impact of information and technology on sustainability. She is also research active in the field of knowledge management with a particular emphasis on critical success factors for knowledge management as well as the role of key stakeholders in knowledge management activities. Mouzughi has been a keynote speaker and invited guest speaker at national and international conferences and has various publications in the knowledge management field, the latest being a research monograph "Critical Success Factors for Knowledge Management" (Lambert Publishing, Germany, 2012). Being of Libyan background, Mouzughi also has research interests in various aspects of Libya's development. Prior to her career in academia, Mouzughi worked in the insurance industry as a project manager for over seven years.

Mark Mulligan is reader in Geography at King's College London and senior fellow at UNEP World Conservation Monitoring Centre. His research focuses on spatial modelling of the impacts of climate, land use change and land management upon water resources, ecosystem services, agricultural production and biodiversity. Mulligan works on the ground throughout Latin America but has also carried out research throughout Africa and Asia. He is developer of the Water World Policy Support System.

Turki Faisal Al Rasheed is the founder of Golden Grass, Inc., an agricultural and contracting company, in 1982. The company has since become one of the top 100 companies in Saudi Arabia. As a sustainable agricultural development expert and the author of five books, including most recently Post Arab Spring, and numerous articles on national security, sustainable agriculture, food security, sustainable development, political economy, and self-improvement, Al Rasheed is frequently requested to speak about these topics in a variety of different settings. He is currently an Adjunct Professor at the University of Arizona, USA, and Visiting Research Fellow at Liverpool John Moores University, UK.

Kassem El-Saddik is an independent researcher and program evaluation expert with more than 10 years of experience in designing, developing and deploying monitoring and evaluation (M&E) systems. He was member of the Policy and Strategy team who managed the Dubai Strategic Plan 2015 and planned for Dubai Plan 2021. El-Saddik led various policy initiatives given his wide knowledge of the development and environment landscape in the MENA and the Arabian Gulf (GCC). He has been actively engaged in various UNEP-led regional and global lobbying and negotiation meetings around environmental institutional reform and the post 2015 development agenda, and served on UNEP Advisory Group on IEG. El-Saddik's areas of interest are environmental policy, governance and sustainable consumption and production. He is a genuine civil society advocate and practitioner. El-Saddik is founder of the Lebanese Evaluation society and a vice president for the Arab Center for Development and Environment in Lebanon.

Kate Schreckenberg is a forest governance researcher with a particular interest in improving the livelihoods farmers obtain from managing their natural resources individually or communally. She tends to work in interdisciplinary projects and specialises in the use of participatory research methods.